The
SUGAR HACIENDA *of the*
MARQUESES DEL VALLE

The
SUGAR HACIENDA *of the*
MARQUESES DEL VALLE

WARD BARRETT

UNIVERSITY OF MINNESOTA PRESS, Minneapolis

Library of Congress Catalog Card Number: 74-110146

ISBN 0-8166-0565-3

The three illustrations of eighteenth-century sugar mills which appear between pages 56 and 57 were taken from Diderot's *Encyclopédie*, Volume 1 (Paris, 1762), in the holding of the Wilson Library, University of Minnesota.

Preface

MANY years ago, while reading Sandoval's work *La industria del azúcar en Nueva España*, it became apparent to me that the continuity of his account depended heavily on the records of a single mill established by Hernán Cortés about 1535. I had just finished a short essay, "Caribbean Sugar-Production Standards in the Seventeenth and Eighteenth Centuries," and in the course of doing the research on which it was based I had been impressed by the lack of published information concerning Spanish colonial plantations. Not only were no handbooks available, of the sort published in plenty in France and England, and which describe both plantation management and the general state of the industry, but neither were there any summaries of individual plantations in New Spain. Since Sandoval's notes suggested that perhaps the Cortés records were full enough to yield the kind of information that eventually might contribute to a comparative study of sugar plantations, I decided to investigate. As it turned out, the records are more than adequate in some respects and deficient in others.

As I became familiar with the records of other mills in the Cuernavaca region, it also became apparent to me that the Cortés mill was anomalous in some important senses, but not from the point of view of technology and management. The plantations of Morelos differed little in these respects, so I used the records primarily to see how these plantations were run, and what

the results of the collective, cumulative experience and insight of the persons associated with this plantation had been. The records did show, however, that the Cortés plantation differed markedly from others in respect of the continuity of ownership it experienced, since for nearly four centuries — from about 1535 until this century — it remained in the hands of the heirs and descendents of the Conqueror of New Spain. By contrast, other mills exhibited considerable turnover in ownership, excepting mills owned by religious orders; none of the latter, however, were started so early as the Cortés mill. Furthermore, this plantation was never encumbered by the mortgages and obligations that burdened most of the rest of the sugar haciendas of New Spain, a feature that began to be important in the seventeenth century.

Since my major aim, after discovering what kind of information the Cortés records could be made to yield, was to emphasize the technological and managerial aspects of the plantation, its special qualities did not loom very important apart from the fact that they had led to the collection of an abundance of accessible records. I decided early in the course of the investigation to pay little attention in this work to the problems that beset other mills and even the industry itself, leaving these matters to a future work that will deal with the historical geography of the industry in Morelos. In addition, I decided to rely on the fact that Sandoval's work described fairly fully the regulations pertaining to the industry, if not their effects, leaving these as well to a later work. These official regulations were not numerous, in any case; only two, the abolishment of Indian repartimientos for sugar plantations about 1600, and the prohibition of the sale of molasses to Indians, seemed important to the Cortés mill, and both of these have been dealt with below.

In respect of sources, I have largely limited myself

to the documents themselves, with the result that the bibliography of printed works is short and restricted to those referred to in the text. Although I have consulted other works that were helpful in various ways, it did not seem to me necessary to list all of them.

Part of this research was supported by small grants from the American Philosophical Society and the Graduate School of the University of Minnesota, which covered some of the costs of microfilm and field study in Morelos. In addition, I am indebted to Mrs. Patricia Kangel Burwood, of the Cartographic Laboratory of the Department of Geography at the University of Minnesota, for accepting my very rough sketches of maps and graphs and producing clear and pleasing results. The director, Señor J. I. Rubio Mañé, and personnel of the National Archive of Mexico have been unfailingly helpful. Woodrow Borah and Howard Cline were kind enough to encourage me in the initial stages of this work, and I want to thank them for doing so. The careful editing of the University of Minnesota Press has greatly improved this book.

W. J. B.

Minneapolis, July 1969

Table of Contents

LIST OF ILLUSTRATIONS

The
SUGAR HACIENDA *of the*
MARQUESES DEL VALLE

I

Introduction

THE history of European settlement of many subtropical and tropical coasts and islands is intimately linked to the history of the cane sugar industry. Together with the search for precious metals and the establishment of trading posts, the manufacture and sale of sugar and molasses, and the labor needs of the industry assisted in forming the elaborate network of commercial, governmental, and intercolonial relationships that was European imperialism. The dependence of the industry on Negro slave labor in many places led to ever stronger ties among Europe, Africa, and the colonies, as well as to one of the largest migrations of people in the history of the world. The relations between sugar producers in the colonies and merchants and refiners in the metropolitan markets affected the politics of the mother countries, for the interests of the latter two groups made them representatives, often influential, of the diverse interests of the colonial producers. To their influence and power in some European countries should be added the influence of religious orders; some, like the Jesuits, were owners of numerous sugar plantations in Spanish America and the French colonies; others had considerable economic interest in the industry through mortgages and other liens on plantation income. Colonial officials were themselves, in many cases, owners of plantations. The links among strategic and military interests and sugar colonies were emphasized as late as 1841 by Daubrée, who supplemented the axiom, "With-

out colonies, no navy," with another of his own invention, "Without sugar, no colonies" (p. 6).

The establishment of sugar plantations followed the occupation of islands by European countries as they thrust west and south in the fifteenth and early sixteenth centuries. After the occupation of the Azores, Madeira, the Canaries, the Cape Verde Islands, and São Thomé, settlers engaged in sugar production in those places for at least awhile. Columbus introduced cane to Hispaniola, and the plantations of his relatives were centers of social life in the early history of the colony; settlers there turned both to sugar production and livestock raising, making hides and sugar the mainstays of its economy for some decades in the sixteenth century. The decline of the industry on most of these small islands, in spite of royal encouragement in the case of the Antilles, was manifest by the late sixteenth century, and in part because of imports of large amounts of sugar from coastal Brazil, cheaper in Europe in spite of the long haul (Mauro, pp. 183, 188), as well as to diversion of Spanish interest from the Antilles to the mainland.

France, England, and the Netherlands began to occupy the lesser Antilles and the Guyanas shortly after the decline of the Spanish Antillean colonies and the rise to pre-eminence of the Brazilian industry. After the acquisition of the larger colonies of Jamaica and Saint-Domingue by England and France, these two countries became firmly established as Caribbean powers, and in the eighteenth century both colonies developed into major producers, as did Surinam for the Dutch. Denmark entered the Caribbean in a minor way as possessor of a sugar colony, and their acquisition ended on a massive scale at the close of the nineteenth century, when the United States acquired Hawaii and virtual or complete control of fragments of the Spanish Empire, all sugar producers: Puerto Rico,

Cuba, and the Philippines, later joined by Santo Domingo.

Besides those colonies and regions whose production was destined primarily for export, there were others that produced considerable amounts of sugar for internal consumption by the early seventeenth century, for the growing and processing of cane early became widespread in the New World. Peru and New Spain are good examples of this kind of colony. Chaunu has shown that, even as late as the period 1568–1620, both the weight and value of sugar exported from New Spain were very small in contrast to exports from Hispaniola and even Puerto Rico, and also in contrast to the value of exports of dyes and hides from New Spain itself (1956, Vol. 6, Pt. 2, pp. 1004–11; 1957, Vol. 6, Pt. 7, pp. 140–43). Probably, also, exports were small in proportion to total Mexican production, for Chevalier estimates that the fifty or sixty largest mills produced annually in the early seventeenth century between three and five thousand tons, in addition to the amount produced by many small trapiches (1956, p. 67). Furthermore, it is clear that the number of mills and the volume of production increased in the same century, particularly after growers extended the cultivation of cane to Tierra Adentro.

Partly because of their small export trade in sugar, the Peruvian and the Mexican industries have received less attention than those oriented toward export; after all, historical writing about colonies has been dominated by citizens of the mother countries rather than of the colonies and ex-colonies, and the political and economic relationships between mother country and colony have been emphasized even by residents of the latter. In the case of the sugar industry of New Spain, the result has been to neglect a very important industry with wide social and geographic implications. Chevalier (*ibid.*, p. 68) estimates that by the late seventeenth century it was more important in terms of capital investment than was commercial wheat growing, and it is a commonplace that several Mexican regions were long dominated by sugar haciendas.

The intimate association of sugar production for export with islands and coasts, where water transport and ease of access to the sea were of crucial importance, led to clustering of plantations and a focus on ports in many places. This kind of pattern is well shown on Mauro's (*op. cit.*, pp. 184, 198), maps of Madeira and Bahia, as well as by the distribution of mills on the southern coastal plain of Santo Domingo. The distribution of the plantations of New Spain was different, because the markets were internal and the pattern of coastal location did not develop strongly there in the colonial period. High transportation costs from the major producing areas to Vera Cruz made export unprofitable; as late as 1823 an anonymous author of a pamphlet bound in L219 E2* pointed out that bad roads and long distances meant that transport costs to Vera Cruz, even from the nearest producing districts, had long been four times those of the province of Havana.

The distribution of sugar producing areas that were developed in New Spain is shown in Figure 1, where the principal regions have been taken from Chevalier's map (1956). Clusters of mills and plantations are emphasized owing to the small scale of the map; small areas served by small mills are omitted for the same reason. Even the major producing areas were small, many were irrigated, and the principal feature of their distribution was their occupation of subtropical basins adjacent to the escarpment of the cooler volcanic plateau and its cities and towns. The major producing area, the Cuernavaca-Cuautla basin, supplied Mexico City, whereas the plantations of Atlixco and the area south of it supplied Puebla and the capital as well; smaller and more isolated areas in Michoacán supplied Morelia and, at least in the seventeenth century, the mines as far north as Zacatecas. Near Oaxaca was another small cluster. In contrast to this clustered pattern developed a second pattern, one of small, scattered areas, perhaps with a single mill or a few trapiches, serving small areas, probably with panocha rather than clayed sugar and perhaps also with molasses.

The Cuernavaca-Cuautla region was the most important in colonial Mexico, since it was nearest to the largest market. By 1600, twelve mills of varying sizes were operating in the region, and of these the Cortés mill was the largest single producer. By 1900, some thirty sugar plantations claimed nearly all of the land of Morelos, and probably the total number in operation since 1600, including those abandoned, was fifty or sixty. After the Revolution, nearly all of them ceased operations, and finally a large cooperative ejido mill was built at Zacatepec. Today more cane is raised and more sugar produced in Morelos than ever before; the Revolution did not free the region from dependence on sugar, but rice and specialty crops have added more diversity. Morelos did not, therefore, pass through the

*The form of the citations is explained at the beginning of the bibliography (p. 137).

4

"cycles" or cyclic stages of development and decline that are so easily accepted by Chaunu and Mauro; instead, like many other sugar-producing areas, it has experienced long-term, persistent, and dominant focus on a single commercial crop.

From the point of view of ownership, the mills were of several kinds. The largest number consisted of privately owned plantations that seem not to have had much continuity of ownership, owing apparently to numerous encumbrances in the form of capellanías and various kinds of mortgages, many incurred in the seventeenth century. The other, smaller, group of plantations experienced much greater continuity of ownership. It consisted of mills owned by religious orders and colegios, most with headquarters in Mexico City, and entailed plantations, of which there may have been only one, the subject of this work. Most of this small group were clustered near Cuautla and Yautepec.

I have dealt with the records of the Cortés plantation in three principal ways: as representative of the history of the sugar industry in Mexico; as representative of the history, external relationships, structure, and management of Spanish colonial plantations; and as a chapter in the history of sugar technology. Fundamental to this work is acceptance of the enormous population decline and its subsequent recovery that occurred after the Conquest of Mexico; probably few important historical and geographical problems posed by Mexican data can be dealt with adequately unless attention is paid to the population changes described by Borah and Cook (1963). The history of the Cortés plantation affords a good example of their effects. In the most general sense, these consisted of opportunities to take over abandoned land formerly owned by a dense Indian population, and a severe reduction in the labor force, resulting in heavy reliance on Negro slave labor in the case of many sugar plantations.

The work of Sandoval suffers from a lack of appreciation of the magnitude and effects on land tenure of these demographic changes, since it long predates the work of Borah and Cook; to a lesser extent, so does the work of Chevalier, although the latter recognized that many haciendas had established their boundaries by the time the population had reached its nadir in the mid-seventeenth century, and that the establishment and growth of haciendas was related to population loss. The later work of Gibson is based on acceptance of the great importance for agriculture in the Valley of Mexico of the changes in the numbers of Indians and associated changes in Indian communality and power. The Cortés plantation, shaped as a unit from several contiguous blocks of land with titles of varying worth, offers a good example of how the population decline affected the formation of a particular hacienda; even the date — about 1642 — when the mill was shifted from its original and finally disadvantageous site to its present one is nearly coincident with the time of lowest population in Morelos.

As population increased in the eighteenth and nineteenth centuries, so also did pressure for recognition of the pueblos' need for more land, which could be gained only at the expense of the haciendas. Owing to their rival claims, the two groups shared a long tradition of violence extending back at least as far as the early seventeenth century in Morelos, and most villages had been unsuccessful in defending their claims against the hacendados. In the early twentieth century, the pueblo inhabitants of Morelos affiliated themselves with the revolutionary forces. Their ideology found its best-known exponent in Emiliano Zapata, who called for a return to a tradition that held the pueblo to be the landholding unit of primary legitimacy, and after the Revolution most of the land of the state was converted to ejidos and small properties.

The effects of population change on the nature and organization of the labor supply in the Valley of Mexico have been described by Gibson, many of whose conclusions are applicable to Morelos and the Cortés plantation, except for the much greater importance of Negro slave labor in the sugar industry. Lavish use of human labor was a major feature of the sixteenth-century plantation, but adjustment to a declining labor force occurred in the latter part of the sixteenth and early seventeenth centuries. As the population increased, a new element appeared that was to be of great importance for the plantations: a racially mixed group descended from Indians, Negroes, and Spaniards, who formed a major and distinctive part of a growing labor pool that consisted also of village Indians and Negro and mulatto slaves. By 1800, the sites of many mills were occupied by sizeable settlements of slaves and free mestizos; by 1900, free mestizos dominated these settlements, which were generally much larger than they had been a century before. These settlements, different in organization and appearance from the Indian pueblos, were necessary to the functioning of the plantations for about two centuries.

Research on the structure, management, and external and internal relationships of colonial Mexican plantations has been inadequate; at the practical level, no one could claim, after reading most of the works dealing with haciendas, to be able to manage such an enterprise. At another level, it is obvious that much writing about haciendas has been, consciously or not, designed to serve the goal of presenting them as expressions of a cruel and merciless authoritarianism, demonstrations of the truth of the Black Legend. Another, related difficulty posed by much research is due to the too easy assumption that they were derived from feudal models, which often leads to the conclusion that they were meant to enhance the prestige of the owner rather than to serve as sources of income.

A major problem posed by exaggerated emphasis on the feudal, authoritarian, autonomous nature of haciendas has been to turn attention inward, diverting it from their external relationships, except insofar as the latter touched on control of the local labor supply; this emphasis has not, however, led to adequate structural analysis. All of the plantations of Morelos had to sell their sugars and molasses, yet it is very difficult to recontruct from published sources the relationships of hacendados and lessees of haciendas, themselves two very distinct groups, with merchants in Mexico City and convents and colegios in the capital and in Puebla, even though these latter groups were major sources of cash necessary to finance daily operations and capital improvements. It is difficult to tell what classes of persons could afford to buy the output and the uses to which it was put. This work does not attempt to answer all the questions posed above, largely because the selling arrangements of the Cortés plantation were relatively simple and the Cortés Estate was the principal source of its funds. But most other plantations did not enjoy so secure a position, and it is necessary to suppose, on the basis of abundant evidence concerning hacendados' bankruptcies, that relationships between them and merchants as well as ease of access to capital were of crucial importance in the history of the industry.

It is unnecessary to cross the boundaries of the region to see how complex and extensive were the relationships of hacendados and their lessees. They were linked in numerous ways to the pueblos of Morelos, not only because these were sources of labor, but also because they supplied building materials and firewood to the mills, and at least some of the molasses could be disposed of to their inhabitants. This range of relationships makes it inadequate to deal with the pueblos simply in terms of social justice. Inter-hacienda relationships were also important; these existed among neighbors, as well as those that were not adjacent to one another—for example, in cases where they shared the water of the same stream. Litigation over boundary and water rights pitted hacienda against hacienda, as well as hacendado against pueblo, but here again it is too easy to emphasize struggle and overlook the importance of common interests. Hacendados might be close or distant relatives; they might belong to the same cofradía or share the relationship of compadrazgo; they might simply be friends, sharing supplies in times of shortage, administrative assistance, and physical support in case of attack by Indians or slave rebellions. The relationships among hacendados and lessees, particularly their communality and self-recognition as a group with common interests, have never been adequately described.

The technological history of the sugar industry is of great importance for several reasons. Until the rather recent development of petrochemical and other industries in tropical and subtropical regions, it was the only large-scale agricultural activity employing heavy machinery and elaborate transportation equipment in these climatic areas. Cotton gins and coffee and vegetable oil processing plants, for example, and even the large banana plantations of Central America with their railroads and port facilities, do not require the heavy and complex processing machinery found in large sugar mills. Even in the colonial period, the major sugar-producing areas were the only places in tropical and subtropical regions that offered opportunities to become familiar with the problems associated with the installation and maintenance of machinery and equipment that was then relatively complex, as well as with a processing routine that was and remains demanding.

With the development of the beet sugar industry in temperate, industrialized countries, it became possible to transfer many of the technical advances achieved in Europe and North America to less developed regions where cane sugar is produced. Much of the processing equipment had been directly transferred from Europe in colonial times: the waterwheel and windmill are good examples of such transfers. Nevertheless, it appears that the colonial phase of the sugar industry was a period of learning, of experimentation, of assimilating newly observed facts, in a part of the world itself newly observed. While reminiscing about his stay in Barbados from 1647 to 1650, Richard Ligon remarked on the rapidity with which the settlers solved a great many problems in the cultivation and processing of this unfamiliar tropical plant, after starting with little more than the plant itself and some sketchy information from Brazil. They scarcely knew what questions to ask, and it was not until they had made numerous trials, committed many errors, and asked more informed questions that they achieved satisfactory results. Indeed, Ligon was so satisfied with the results that he was able to assert that he could conceive of no further or necessary improvements.

It is difficult to assume that the industry failed to undergo changes that expressed heightened insight, yet an interesting feature of seventeenth- and eighteenth-century handbooks is the fact that they suggest little change in technology, efficiency, and contemporaneous assessments of the optimum size of plantations between the time the industry began in the French and British islands until the introduction of steam power and the centrifugal process. This conclusion, which I reached (1965, p. 168) after inspection of numerous French and English handbooks, remained to be tested by inspection of the records of an actual mill, and the Cortés plantation seemed ideal for the purpose, since it had the longest history of any mill in New Spain until it ceased operation in this century. This work emphasizes that a combination of factors, in spite of little technological change or total area cultivated, did indeed operate to produce increases in production and efficiency in the colonial period.

This aspect of plantations is related to the third important set of questions dealt with here, concerning the interrelationships, structure, and management of plantations in Mexico. The relationships of planters and their employees within relatively compact regions such as Morelos and Bahia in Brazil obviously deserve a great deal of attention, not only in respect of the diffusion of technology and innovations, but also in respect of their self-sufficiency. Uniformity of technology and procedure implies knowledge shared through relatively frequent contact; self-sufficiency is often taken to mean relative isolation. Chevalier (1956, pp. 69–70), writing of sugar plantations in Mexico, speaks of their being "practically self-sufficient," their tendency "to withdraw . . . behind their own boundaries" in the administration of justice, their nature as the "first great feudal estates" that anticipated the "classical Mexican hacienda." His emphasis is on relative isolation, autonomy, self-sufficiency, yet none of these terms is strictly applicable to Morelos. Probably few plantation owners in Morelos, apart from religious foundations, were exempt from the rapaciousness of some of the Governors of the Cortés Estate and their lieutenants; at least by the eighteenth century, slaves could and did institute suits against mayordomos. Thus, in neither their external nor their internal relations were they autonomous, not even in the administration of justice. As far as the Cortés plantation is concerned, self-sufficiency was neither achieved nor expected: its object was to produce sugar, molasses, and a profit and, practically speaking, it produced little else besides financial losses. Nor was it isolated: all of its inventories, as well as the inventories of other plantations that I have seen, show that personnel of other haciendas, skilled in all the branches of operation, were present as assessors. Inventories offered only one way in which numerous persons could become familiar with the equipment and mode of operation of other haciendas in the region, for, in addition, skilled and managerial personnel held positions at numerous haciendas during their careers, besides moving into and from the region.

We are left with a picture of Morelos rather similar to that of Madeira, Barbados, Saint-Domingue, Jamaica, and Bahia: sugar plantations that were in compact, well-defined regions with a fairly high density of mills were parts of an intricate system of relationships — commercial, institutional, family — with well-developed and frequently used lines of communication. The mill sites of Morelos, like the others, were places of great local importance, to and from which flowed people, their knowledge and ideas, goods, livestock, and cash. Their principal extra-regional focus was Mexico City, where most of the sugar was sold, and where the agents, creditors, and debtors of the owners lived. In addition, most of the mills of Morelos were linked, in varying degrees, as parts of the Estado del Marquesado del Valle, the estate which was entailed by Cortés about the time that he founded his sugar mill in Cuernavaca.

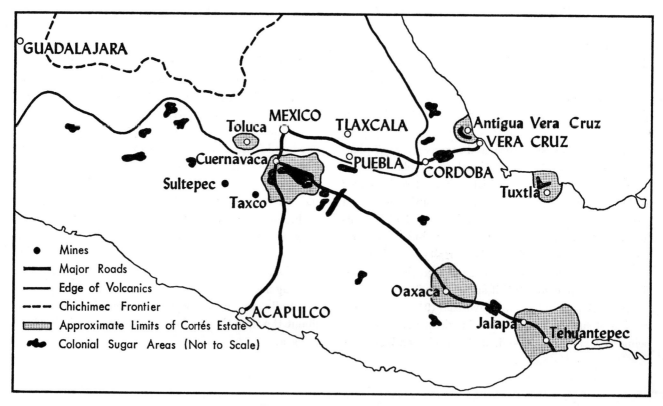

Figure 1. The Cortés Estate

II

The Cortés Estate

FOR a time, the properties entailed by Hernán Cortés constituted one of the largest estates in the New World. The purpose of this chapter is to describe the extent and nature of this group of properties and rights and the way in which they were managed, since without some knowledge concerning them it is difficult to understand certain features of the Cortés plantation. I have selected mainly information that serves to show the importance of the plantation within the larger complex, how it was related in other ways to the administration of the Estate, and how events that affected the Estate also affected the plantation.

Formation of the Cortés Estate

Many of the properties and rights acquired by Cortés continued until the nineteenth century as an entailed unit called the Estado y Marquesado del Valle. Probably the best known were the seignorial rights pertaining to twenty-two Indian towns granted in a famous Real Cédula of 6 July 1529 (Hernández Sánchez-Barba, pp. 596–99). In another proclamation of the same date, the king granted Cortés the title of Marqués del Valle de Oaxaca and, a few days later, power to entail whatever properties he wished (*ibid.*, pp. 599–600, 541–45).

The distribution of the entailed and other properties claimed by Cortés is shown in Figure 1, based largely on data in his will of 1547 (*ibid.*, pp. 554–77). The area

north of the Chichimec frontier held little interset for Cortés, whose attention instead focused on well-settled areas and potential ports, leading him to lay claim to extensive areas on both sides of the Isthmus of Tehuantepec, the Valley of Oaxaca, and what is now the state of Morelos. Mexico City itself was ringed by areas either owned by the Estate or paying tribute to the Estate, including the basin of Toluca and land in Coyoacán. The arrangement of these holdings, focusing as they do on Tehuantepec, Mexico City, and major pre-Conquest trade routes, suggests that Cortés rapidly acquired insight into the geography of Indian Mexico.

Sources of Income of the Estado del Valle

Tribute Payments. The basis of sixteenth-century prosperity in New Spain was the large and dense population of Central and Southern Mexico. Cortés was quick to claim large numbers of Indians as tribute payers, and their importance as a source of revenue to the Estate can be seen from Table 1 (p. 127), which shows that by 1570 nearly half of Estate income was derived from tribute payments. Their magnitude decreased with population in the sixteenth and seventeenth centuries.

The changes of colonial population in Morelos can be described only by presenting changes in the number of tributaries; a review of the problems associated with attempting to convert numbers of tributaries to total population is given by Borah and Cooke (1963), and the major sources on which the following discussion is based are described in Appendix A (p. 109). A principal problem is the establishment of a ratio between tributaries and non-tributaries. Another important problem derives from the likelihood that the total population and numbers of tributaries changed at a different pace, with the latter lagging behind the former owing to the inability (or unwillingness in periods of popula-

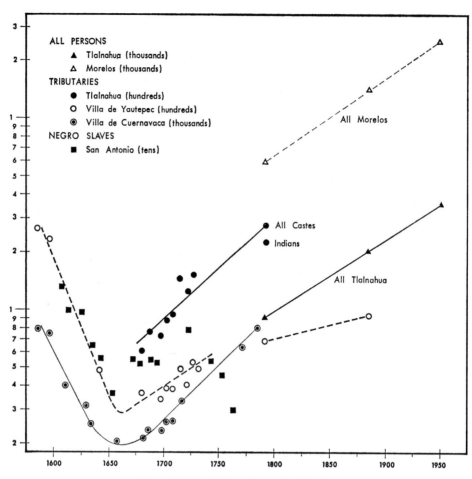

Figure 2. Population change in Morelos

tion decline) of the taxing authorities to keep abreast of population changes.

Given these difficulties, I present in Figure 2 curves showing changes in numbers of tributaries, as well as curves for later dates showing changes in the total population of Morelos; the order of magnitude is indicated in each case by the location of most of the symbols. These curves present separately data concerning individual towns and groups of towns in Morelos, and my analysis will be restricted to examining a composite trend based on these discontinuous lines. It is obvious that the population trends are in remarkable agreement, including even the total number of slaves at the Cortés sugar plantation, which followed a course of decline in the early seventeenth century very similar to that of the nearby and important Villa de Yautepec. Although only the Yautepec and Cuernavaca curves cover more or less adequately the period both of decline and gradual increase in the late seventeenth century, the parallelism between these relatively complete curves and the other

segments is sufficient to justify the conclusion that the various parts of Morelos shared a common population history of decimation, owing probably to introduced and indigenous epidemic disease, followed by increase to the present, with a common inflection point near the middle of the seventeenth century. It is interesting to see that the population was never stable in an absolute sense, but rather experienced long-continued trends of decline or increase; and it is also important to note that there is no point on these curves where the influence of modern public health is yet visible. Probably for most of the population of the region, modern medicine has been introduced on a modest scale only since about 1960.

It may be assumed that the contribution to Estate income from tribute payments followed a course similar to that of population. I have not analyzed many years of Estate income to establish the truth of this assumption, but Table 1 shows that in 1681 tributes were only a third of their size a century before, whereas by 1807

10

they had nearly recovered to the level of tributes in the 1570's.

Sugar Manufacture. This industry was begun in New Spain either by Cortés at Tuxtla about 1528 or 1529, or by Antonio Serrano de Cardona near Cuernavaca about 1530 (L282 E4; L277 E4). The third mill in New Spain was built by Cortés adjacent to Serrano's plantation, in the village called Tlaltenango. The Cuernavaca plantation made a fairly substantial contribution to Estate income in the sixteenth century, but much more modest contributions in the seventeenth and eighteenth centuries, when it was run at a loss for various periods. During the disturbance of Independence and the following several decades, the plantation was also not very successful. Not until the California gold strike created a large but temporary demand for Morelos sugar did the plantation become extremely profitable, contributing probably almost half of total income of the remnants of the Estate for the years 1850–53.

The plantation at San Andrés Tuxtla was operated for a little more than fifty years, from about 1530 to about 1583, and I have found no annual records of its activities. It is worth noting that the Tuxtla mill was much more advantageously sited for shipping sugar to Spain than was Tlaltenango, but less so for shipping to Mexico City. Possibly the partial revival of the sugar industry in Spain in the late sixteenth century made the Tuxtla mill uneconomic, and it was also affected by decline of the Indian population in the region.

Livestock. The raising of livestock was extremely important to the Estate. Cattle and mules were raised at Jalapa (del Marqués, in Tehuantepec) in a tropical lowland with fairly abundant water supplies, whence they were driven via Oaxaca to Cuernavaca, where Cortés had a large estancia called Mazatepec about twenty-five kilometers southwest of the city. The city of Cuernavaca as well as the mines and the sugar plantation were supplied with livestock from Mazatepec, and cattle for the slaughterhouses in the Valley of Oaxaca also came from Jalapa. A tannery belonging to the Marqués produced leather, some of which was exported to Peru (Chevalier, 1956, p. 171). Horses were raised at a brood mare ranch at Tlaltizapán, Morelos, and also at Matlultizinco, Oaxaca; the former produced mules as well (Archivo General de la Nación, *Documentos Inéditos Relativos a Hernán Cortés*, pp. 288–89).

Sheep and Obrajes. There was an obraje at Cuernavaca adjacent to Tlaltenango. However, the fulling mill (batán) was described in the inventory of 1576* as "podrido."

Silkworm Raising. In the 1540's Cortés instituted silkworm raising on a large scale at Coyoacán in the Valley of Mexico and at various places in Morelos, where large mulberry plantations were established in 1544, a large rearing house built in Yautepec, and the first raising occurred in 1546. Borah's (1943) study shows that the industry had a short history in the Marquesado, since after Cortés' death the trees in Morelos were on the verge of extinction, and this rather large-scale experiment failed shortly thereafter.

Mining. Gold mining was an activity of minor importance to the Estate in the early part of the sixteenth century, and only near Tehuantepec, where Cortés opened several mines after about 1538 (Cadenhead, pp. 283–84). Before the opening of these mines tributes had been paid in gold by some of the Indians. An epidemic in 1545 greatly reduced the number of Indians available to mine, with the result that production declined. However, according to Alamán (*Disertaciones*, Vol. 2, p. 73), some placer mining continued in Tehuantepec until about 1643.

The mining of silver probably produced greater benefit for the Estate, and some mines opened by Cortés at Taxco, Sultepec, and Zumpango (Colima) in the 1530's continued operating into the seventeenth century. Provisioning the mines was an important function of the agricultural areas of Morelos, whence wheat, corn, and livestock were sent to them. Sometime before 1639 silver mining ceased to be important to the Estate; in a review of expected income drawn up in that year, it was noted that the Sultepec mines had been abandoned, and that the Taxco mines had been rented at three hundred pesos per year, but even this small rent had not been paid (L239 E21).

Censos. In addition to income from the activities mentioned above, the Estate received variable income from censos, an annual rental payment whose amount was up to 5 per cent of the value of the land leased. Effectively, censos were perpetual leases of land, water, and mill rights to which the heirs of the lessee had first option. Estate censos were especially numerous in Morelos, where many sugar plantations and ranches had

* Where it is clear that information has been taken from inventories or accounts of the plantation, reference to the documents containing the latter will not be given in the text; the reader may refer to the bibliography for the location, by year, of all inventories or annual accounts.

their origins in early seventeenth-century grants made by Pedro Cortés in exchange for annual payments that were little more than nominal.

Mexico City Rental Properties. The Estate and the Hospital de Jesús rented lots and houses in Mexico City. Prophetically, Cortés stated in his will that he expected this urban property to become valuable, and finally it did become the Estate's single most important source of income. Most of the property was very near the Zócalo—for example, some houses belonging to the Estate and to the Hospital were just north of the Cathedral on what is now the Avenida Guatemala, and hence they were valuable for their location (*ibid.,* p. 196).

Trade with Peru. The story of Cortés' lack of success in this venture has been reviewed by Borah (1954), who shows that his interest had resulted in the establishment by 1540 of the largest and best-equipped shipyard in New Spain. Interest in Peru lasted until the 1550's, when a final trading expedition was sent with some sugar from Tlaltenango, but like the trading ventures of the first Marqués it was a failure. Perhaps the opinion written in 1539 by one of his agents to Cortés concerning the lack of success in the Peruvian trade should have received more attention than it did: "Siempre he conocido que no nació V. Señoría para mercader" (Alamán, p. 67).

Changes in the Composition and Magnitude of Estate Income

Some of the important changes in Estate income are shown in Table 1. Although the abundant data in the Hospital de Jesús collection permit a much fuller analysis of these important changes, I have selected a few of the most easily analyzed years in order to indicate their general nature and the relation of revenue from the sugar plantation to gross Estate income.

Average annual income in the late 1560's and the 1570's is probably fairly well illustrated by a composite set (under ca. 1570), and best discussed by contrasting it with other years. The major contribution to Estate income, in 1570 as in 1807, was tributes, which accounted for about 40 to 45 per cent of the total in both years, if one adds to the tributes the Crown pension, *Recompensa por la Villa de Tehuantepec* (compensation for tributes lost by the Estate to the Crown). In addition, it should be noted that total gross income was approximately equal in both years. About $20,000* of the gross in 1570 was derived from rural properties, of

which the single most important was Tlaltenango, then earning an annual rental of $9,000.

The effects of population decrease and economic decline stand forth clearly in the period between 1570 and 1807. By 1681 tribute income had shrunk greatly, and decline was obvious in the agricultural properties as well. Ninety years later, in 1770, tributes had increased but there was little change in agricultural income, and even the Mexico City properties returned less. As in 1681, about 10 per cent of total gross income came in the form of cash and grain payments from Chalco compensating for the loss of Tehuantepec tributes.

The representativeness of the data of 1681 can be assessed by reference to Tables 2 and 3 (p. 123). Table 2 shows for various years before 1671 the amount of money that the Governors of the Estate had contracted to send to Spain, and for some years the amount actually sent. The first major readjustment to population decline occurred about 1645, when the Governor appealed for revision of the amount expected from tributes on the grounds that depopulation made the total contracted amount ($51,000) unrealistic (L237 E17). Other years (1640, 1643, and 1647) falling within the same rental period show, by the very small amount of money sent to Spain, that the difficulties of collecting tributes and probably other amounts owing had become very great. For more than twenty years (1649–71) the amount contracted to be sent was about $38,000, instead of the earlier $51,000, and the actual amount sent in one of those years (1671) was less than half the amount contracted.

The earlier shipments to Spain are comparable with the column headed "Balance" in Table 3, showing net income after deduction of expenses of collection and administration for the years 1686–1704. Average gross income ($50,917) differed little from expected gross income in 1681 (Table 1: $58,083); deductions constituted, on the average, about 20 per cent of the gross, and the balance ($40,647) differed little from the amount contracted to be sent to Spain earlier in the seventeenth century (ca. $38,000).

Recovery to the levels of the 1570's had occurred by at least 1807. The change in composition of income between these two times is best illustrated by the accompanying tabulation, containing approximate values in a reduced number of categories. Practically all of the

*Throughout this work, the symbol $ refers only to either the sixteenth-century peso de oro común or, later, the peso. A sum of money such as $6-2-5 should be read as six pesos, two reales, and five granos.

change was due to changes in income from censos and urban properties, neither of which was important in the sixteenth century. Although none of the censos amounted to a large sum, and many were difficult to collect, the large number granted meant a fairly large total income from them by 1807. The level of income from rural property was due largely to the improved performance of the sugar plantation, since other major rural properties declined in importance after the sixteenth century. The increase in the income from urban properties refers to those in Mexico City.

	ca. 1570	1807
Tributes	$50,000	50,000
Rural properties	20,000	24,000
Urban properties	3,000	34,000
Censos	none	17,000

In the sixteenth century, net Estate income had been fairly large in comparison with shipments of precious metals to the Crown. Gómez de Cervantes (pp. 187–89) presents a tabulation of the value of Crown shipments from 1522 to 1601, showing that their value in 1570 was slightly more than $800,000, when Estate income was in excess of $100,000. However, during much of the seventeenth and eighteenth centuries the net income of the Estate appears to have been under $50,000, half the total of 1570.

Changes in the Administration of the Estate

Besides being affected by economic and demographic trends felt throughout Mexico, the Estate was also greatly affected by events more or less special to itself. These included the death of the founder, Hernán Cortés, in 1547; the seizure of the Estate by the Crown in 1567, after which it was supervised by Crown officials until 1574, when it was returned to Martín Cortés; and the rather long bankruptcy proceedings against Fernando Cortés in the early seventeenth century. The latter event was followed by a long period of routine administration, succeeded by what appear to have been structural, administrative, and procedural improvements in the eighteenth century. The Napoleonic Wars and Mexican Independence, accompanied and followed by economic and political instability in Mexico and Europe, resulted in decades of uncertainty that culminated in the effort, begun in 1835, to break up the Estate and sell its parts. All of these events had lasting effects on the structure and operations of the Estate, affecting as well the plantation at Cuernavaca.

Death of Hernán Cortés. It is generally agreed that the financial affairs of the Estate were not very sound at the time of Cortés' death, and that the bequests he made were too heavy to be borne by Estate income (Alamán, *Disertaciones*, pp. 73–76; Borah, 1954, p. 38). If, in analyzing the provisions of his will, we separate recurrent annual obligations from those that could have been discharged by one payment (in fact, these were not), it is only the latter that appear sizeable (Table 4, p. 127). Single payments totaled approximately $320,-000, of which about $275,000 was owing on dowries of his three legitimate daughters and his second wife, Doña Juana. The guardians of Martín Cortés were to be paid at a rate of $1,200 per year until he reached the age of twenty-five, a total payment of about $12,000. Annual payments of $16,500 to Martín Cortés were to cease within about five years, after which the entire income of the Estate was to be his; the other pensions were for one or two lives, and might have been expected to end within thirty years. Thus, the only large recurrent payments were those to the Hospital de Jesús and to the Convent and College in Coyoacán, but since the latter were not built as Cortés had directed, the Hospital was the only large charitable commitment in his will. It was to be maintained by the rent from land and houses in Mexico City, and the Estate was liable to annual payments not exceeding $5,500 when the rents fell below this figure.

This substantial charitable gift does not seem large in comparison with the net worth or net income of the Estate, particularly when one considers that a large part of its mid-sixteenth–century income consisted of tributes that may have cost little to collect. The total amount of bequests and annual charges should be compared with gross Estate income in the 1570's and 1580's (L460 E1, E1A, E2, E4) running well over $100,000 and in some years over $200,000; in this light they appear conservative. Depopulation, litigation, and the troubles of Martín Cortés introduced problems unforeseen by the founder of the Estate.

Sequestration of the Estate. Martín Cortés had been in New Spain less than four years when he became involved in activities that appeared treasonable, and he was ordered to return to Spain. In 1567 the Estate was sequestered, and finally he was fined 50,000 ducats and ordered to lend 100,000 ducats to the Crown; in addition, the civil and criminal authority of the Estate was temporarily suspended. According to Alamán (*Disertaciones*, pp. 100, 102), the Estate suffered a great deal at this time, not only because of the fines and the loss of privileges and income, but also because Doña Juana de Zúñiga was still alive and receiving her pension. A Real Cédula of 7 May 1575 provided that Martín Cortés could sell whatever properties of the Marquesado were least profitable to the value of 40,000 ducats, in order to make the forced loan (*ibid.*, p. 187). He sold lots around the Plazuela del Volador in Mexico City, retaining possession only of the Plazuela itself (*ibid.*, p. 192). The fines and loss of income during sequestration may well have been related to the difficulties faced by Fernando Cortés when he assumed the title on the death of his father in 1589.

Bankruptcy of Fernando Cortés. According to an Estate lawyer writing in the 1650's, Fernando Cortés' bankruptcy in 1600 had the effect of giving control of the Estate to the creditors, who rented it to the Marqués; he, in turn, was obliged to rent it to a governor in order to repay them (L295 E137).

The creditors were ranked primarily according to the dates at which the debts were contracted (Sentencia de Graduación, 26 April 1600, L260 E5). The highest priority was assigned to payment of the largest and oldest debt, the dowry of María Cortés; the date of the dowry contract was 13 June 1556, and the amount owing in 1600 was nearly $50,000. Few other debts remained outstanding from the period 1556 to 1586, after which both Fernando Cortés and his brother began to borrow from numerous persons, finding themselves unable to pay either interest or principal from that year on; the borrowing began three years before the death of Martín Cortés. If we subtract from the total debt (roughly $320,000) nearly $50,000 owing on the dowry and about $9,000 owing on two other debts contracted before 1586, it appears that the brothers had borrowed in the twelve years from 1586 to 1597 at an annual rate of $20,000. Since by the late 1590's the unpayable annual interest was about equal to that figure, it could not have been many more years before interest charges alone would have consumed most of the income from the New World.

The creditors were to be paid not only from current Estate income, but also by the sale of personal belongings of the Marqués in New Spain. On 22 November 1604 an inventory was taken of Tlaltenango on the grounds that it was one of the bienes libres or unentailed properties of the Marqués, but Pedro Cortés objected that it was in fact entailed (L260 E1). Probably uncertainty concerning its final disposition, combined with the effect of Viceregal edicts regulating the use of Indian labor on sugar plantations, caused a decline in production at the plantation at this time.

Administrative Structure of the Estate

Direct control by the Marqueses del Valle over the administrative details of the Estate was lost after the concurso de acreedores brought about by the debts of Fernando Cortés. The structure through which they had worked before the concurso changed little, however. Previously, the official directly below the Marqués was usually called mayordomo and charged with general supervision of Estate affairs. His work consisted largely of routine collection and disbursement of funds and materials as well as the conduct of lawsuits; by the 1550's his duties had assumed the character that they were to retain for the rest of the century (well outlined in L243 A2–4). The existing fragments of correspondence between the Marqués and the mayordomos of the Estate show that all of them received detailed and probably frequent instructions from the Marqués; most of the fragments date from the time of Martín Cortés, and show that whatever else he may have done the second Marqués had made himself conversant with the details of his business enterprises.

After the concurso it became customary to lease the administration of the Estate for periods of nine years and for a fixed annual rent to the highest bidder, following advertisement in Spain and New Spain. Apparently only after leasing became customary did the title of the chief administrator change from mayordomo to Governor.

Governor of the Estate. The Governor's full title, Gobernador y Justicia Mayor del Estado y Marquesado del Valle, emphasized the dual aspects of his role. His administrative powers represented a continuation of the major feature of the sixteenth-century position of mayordomo, although his judicial power seemed more prominent in the seventeenth century; by this time, it was generally agreed that loss of the judicial powers would make the position thoroughly unattractive, since a major feature of the position of justicia mayor was the power to appoint local officials who exercised judicial functions (L246 E1).

The rights and duties of the Governor were listed in detail on 2 June 1612 (L258 E1), when Pedro Cortés granted the position to Cristóbal de Molina. The power to collect debts, review and approve accounts of mayordomos, rent units of the Estate, and buy supplies and equipment were of foremost importance. The Governor was also empowered to represent the Marqués in both civil and criminal cases, and in both civil and ecclesiastic courts, as well as to remove and appoint officials, including the offices of corregidores, alcaldes mayores, judges, and other ministros of the Estate and Hospital —in other words, all the positions concerned with administrative, fiscal, and judicial affairs. Ultimate control rested with the Marqués, since only he could appoint the contador mayor and the Governor was to perform orders of the Marqués.

Performance of the Governors. From the point of view of the Marqueses, the primary function of the Governors was to collect the rents and tributes due the Estate. At some time in his career as Governor, each drew up a list of expected annual income against which to compare the amount collected. In addition, the independent controller of the Estate had the same information, and his major task was to compare the amounts due with the amounts collected. Table 5 (p. 128), summarizing the amounts owing by various Governors, is based primarily on controller's statements. If the list were complete, it would probably show that most Governors — especially those who remained in office more than five years or so — from the late sixteenth century until control was assumed by Alamán in 1825 were regarded by the Marqueses as debtors to the Estate upon their retirement from the office.

Since experience showed that nearly every Governor could expect an expensive lawsuit at the end of his tenure, and many had at least some of their possessions confiscated and sold as a result of judgments favorable to the Estate, it seems unlikely that anyone would undertake this work unless it provided opportunities to gain large sums of money in ways that are not obvious in either the court records or the Estate accounts. Probably most of the Governors were swindling both the Estate and the people under its jurisdiction, tempted by the numerous opportunities offered by their appointive and judicial powers; possibly, also, the sale of offices provided a large income.

Estate Controller. The official called contador del Estado had great importance because of his power of review of accounts of the units of the Estate. Numerous separate accounts passed under his scrutiny: tribute accounts from each district, accounts from the plantation and the Hospital. Receipts were required even for transactions involving very small amounts of cash, and the existing bulk of such receipts is an impressive monument to diligence and routine. The controller also supervised the reworking of accounts into a standard form for review by the Marqués, and he had the power to suggest changes in the work of other officials (1575: L208 E5). Controllers were involved in many fewer suits than were Governors.

Estate Lawyer. The principal legal counsel was part of the highest administrative hierarchy of the Estate. He had command of a small staff of scribes, and his major task was to defend the interest of the Marqués in the courts, as well as to initiate litigation after consultation with the Marqués or Governor and controller. Under various titles, this important position persisted from at least 1542 (L257 E14 D100).

Officials shifted easily from one position to another at the higher levels, with the exception that no one filled the more highly specialized position of lawyer without the necessary training and experience. The same pattern of easy transfer within the administrative and supervisory positions is present in the plantation structure, suggesting that there also the routine of each position differed little from most others. Apparently these positions required less a demonstration of skill and versatility than of trustworthiness: a major function in any position was to prevent embezzlement by other officials. Conceptions of growth, development and change seem not to have been present.

It is clear that many officials and employees of the Estate entered its service through the help of relatives already employed; even casual inspection of records quickly establishes this impression, and if the surnames were better guides to relationships, doubtless nepotism would be revealed as the normal means of gaining employment. This feature of the Estate is present from the sixteenth century through the nineteenth; a notable example is the assumption of control of the Marqués' affairs in Mexico by a son of Lucas Alamán after the death of the latter in 1853.

Eighteenth- and Nineteenth-Century Changes in Organization

By the early eighteenth century the administrative structure of the Estate had changed somewhat, expressing the resumption of greater control by the Marqueses del Valle, possibly as a result of satisfaction of the debts that had caused the bankruptcy proceedings. As early as 1721 (T1965 E1) the major officials of the Estate — the Governor, Estate lawyer, controller, procurador, and the administrator of the houses and censos in Mexico City—met together as a group to discuss Estate affairs. In T1965 E1 is an early reference to this kind of group decision-making, wherein the group is described as the Junta de Estado, concerned with approval of a bid to rent San Antonio de Atlacomulco. Much later, in a letter of 12 April 1825 (L219 E3) to the new Governor, Lucas Alamán, the Marqués outlined the relatively simple structure that the Estate had had for more than a century, referring to this same group as the Junta Administrativa. The Apoderado General, a title used thenceforth rather than the older Governador, was to preside over weekly meetings with the other four officers of the Estate, who were the same officials as those described in the document of 1721 with the exception of the procurador, for whom was substituted the escrivano de cámara; the four officers other than the Governor were described as vocales miembros. This nineteenth-century organization appears close to the sixteenth-century pattern, wherein the person holding the Marqués' power of attorney supervised the administration of the Estate for a fixed salary, in marked contrast to the seventeenth- and eighteenth-century pattern of rental of the office of Governor in exchange for a fixed annual shipment of part of the gross income to Spain. There is no evidence that in the sixteenth or seventeenth centuries such a group met regularly, but there is abundant evidence that documents were passed from one official to another, each adding his opinions to the body of the document or making marginal notes.

Another important change of the eighteenth century was the assumption of rather significant powers of review of decisions and events by the agents of the Marqués in Madrid, who were operating through an institution called the Dirección General del Estado del Valle at least by 1747, when Antonio Güell was described as Apoderado y Director General del Estado (L36 V68 E21). It is clear that the decisions of this group took precedence over those of the Junta in Mexico City, and toward the end of the century their links with New Spain became closer when at least some officials went to the colony to assume direct control of operations there (L298 E10; L232 E2; L298 E31).

After 1800 the fortunes of the Estate reflected the political changes occurring both in Mexico and in the Old World as a result of the Napoleonic Wars and the revolution in Mexico that led to Independence. The properties were confiscated by both the royal government and the Mexican government, and finally disentailment began in accordance with a law requiring the breakup of entailed estates.

The fighting in Mexico cost the Estate and the Hospital large sums of money. For example, about 1809, the Viceroy ordered the Governor to give him approximately $400,000 stored in the safe of the Estate, representing the rents of several years from the Marquesado but not sent to Spain because of the war; some of this was repaid (Alamán, *Disertaciones*, Vol. 2, p. 95). The Hospital, whose rents were confiscated by the Crown at the same time, suffered the loss of $45,000.

San Antonio was also affected by the military and political events of the period. The plantation was occupied by the rebels for a period of about four months in 1812, during which time work was suspended. The plantation was again physically threatened in 1821, when Iturbide marched on Cuernavaca in March; shortly after, the plantation was abandoned except for a small group that remained at the mill until October, when field and mill work resumed. Some stock had been lost, as well as more than 1,000 tareas of cane owing to the fact that it had not been milled in time (L219 E2). In 1822 the Marqués was ordered by the government not to sell or rent the plantation, and the Junta faced the problem of restoring the fields and waiting more than a year for a crop without the cash (approximately $40,000) necessary to do so (L219 E2).

It is impossible to estimate the amounts paid and lost by the Estate and the plantation as a result of the Independence movement. Both the viceregal and national governments levied special assessments, and the plantation supported a number of troops for many years, besides suffering occupation and disruption of field work in at least 1812, 1821, and 1822 that lasted long enough to produce loss of cane and income, and to cause reinvestment and delay of income that may have totaled $75,000.

Planning by the Junta continued in spite of these reverses, political change, and the issuing of paper money that by 1822 (L219 E2) made the financing of daily operations very difficult. The price of sugar was low, and it had become difficult by 1822 to obtain labor even with cash advances; by August the plantation administrator reported that workers refused to work at any of the haciendas in the region. The situation became more serious in 1823, when he reported a conspiracy to seize the property of "Europeos" (L219 E3). It was in this setting of uncertainty that Lucas Alamán began in 1825 the association with the Neapolitan Duke of Monteleone and Terranova, the heir of Cortés, that lasted until Alamán's death.

Administration of Lucas Alamán

Alamán's major task for approximately the first decade of his administration was the supervision of Estate affairs in an atmosphere of political hostility that culminated in confiscation of the property of the Duke of Monteleone by the Mexican government in 1831, followed by its release in 1835. For at least part of this period, some of the income of the Estate and the Hospital was assigned to public instruction; the rest went to the government, and its recovery never seemed likely.

Immediately after the properties were released another and larger task faced Alamán: the sale of the entire Estate, in conformity with a new law requiring disentailment. This complex duty involved the sale of the Mexico City real estate, including the Plazuela del Volador; the sale of Atlacomulco, which did not occur; the sale of the Haciendas Marquesanas, as the rural properties in Oaxaca and Tehuantepec were called; the sale of land leased under the censo system; and an attempt to collect rents and debts owing to the Estate, mostly incurred in the previous twenty-five years of disturbance. By the time the dissolution of the Estate had been accomplished, it was Alamán's intention that only the patronage of the Hospital should remain in possession of the Duke, in order to realize the intention of the founder and commemorate his memory.

In view of the numerous difficulties attending the liquidation of the Estate, Alamán did remarkably well.

Money was scarce owing to political uncertainty, and the persons in power were avowed enemies of the Estate (Alamán, *Documentos*, Vol. 4, p. 296), with the result that the few potential buyers in Mexico were concerned lest confiscation follow the purchase of Estate property. It was difficult to fix prices for much of the real estate (*ibid.*, p. 294), and Alamán felt that any sale price would probably be less than his conception of the real value. However, the danger of losing all in a new revolution pushed him toward immediate sale.

Large amounts of money were collected. In 1835 (*ibid.*, p. 270) Alamán estimated that the Duke could be certain of receiving about $500,000 for the properties; by 1839, when there remained unsold only Atlacomulco and uncollected only debts that he claimed the Mexican government owed the Duke, Alamán had sent nearly that amount to Europe (*ibid.*, p. 433). In October 1849 Alamán wrote that he was sending the last amount of money received from sales of property, but by this time instability in Europe had persuaded the Duke that Mexican investments were no less safe than others, and he instructed Alamán to begin buying property again (*ibid.*, p. 513). The change of plan suited Alamán well, for he had written in the previous March (*ibid.*, 498): "Mi empeño es siempre formarle a V. una nueva casa aquí . . . pero siendo tan insegura hoy la suerte de los propietarios en todas partes, casi no hay otro partido, que tener algo en diversas naciones, a ver que es lo que escapa."

Nevertheless, San Antonio de Atlacomulco remained the principal rural property of the descendents of the Conqueror of Mexico.

III

Leasing and Administering the Plantation

THE Estate officials had five choices concerning the sugar plantation. The choice made through most of its history was administration by the Estate, meaning that the Governor or the Marqués appointed the mayordomo and provided the cash and supplies necessary to keep it in operation. Figure 3 shows that for the period 1540–1847 this mode of operating the plantation was in effect about half the time; after 1847, until the mill was abandoned in this century, the plantation continued to be administered in this way. The most important alternative was to lease the plantation to someone willing to assume the burden of arranging for its weekly requirements of cash and supplies. In the period 1540–1847 the plantation was rented about half the time.

Two other alternatives existed: to suspend operations temporarily, which seems to have been done briefly in the early seventeenth century, or to abandon the plantation entirely, as in the second decade of this century. The fact that temporary suspension of operations occupied so little time — less than a year — is unusual, since most of the plantations of Morelos seem to have experienced both more numerous and longer suspensions of operations. The fifth option, that of selling the plantation, was much discussed but not exercised.

This chapter is concerned with the advantages and disadvantages of leasing, the conditions of leases, and the problems of administration by the Estate.

Leasing the Plantation

Characteristics of Leases. Most of the leases were supposed to run for nine years, the first four or five obligatory, the last four or five optional. Inventory was taken at the beginning and end of the period; discrepancies were made up by the renter, or, in cases where the value of his improvements exceeded the discrepancies, the balance was paid by the Estate. In nearly every case a lawsuit or the threat of a lawsuit marred the relations between the parties, some continuing for decades before a settlement was reached.

Leasing was arranged in a way very similar to the rental of the Estate itself, with proclamations announcing its availability and asking for bids from interested persons. In some cases, a prospective lessee might notify the Governor of his interest, and this event might set off a round of proclamations and, perhaps, bidding by others. Bidding was closed publicly after a set of repetitions — perhaps twenty, often thirty — by the public crier of an announcement of the availability of the plantation. Bidders submitted conditions that in most cases were either identical with or very similar to the conditions of the immediately previous contract. The major issue centered on the annual rent. The Governor might refuse all bids on the grounds that they were too low, but in most cases the highest was accepted.

The successful bidder was granted a short period of time to deposit with Estate officials the contracts of his bondsmen, each of whom guaranteed part of the value of the rent or of the plantation itself in case the lessee defaulted. Usually they were not liable for more than $2,000 each. Estate officials had the right to pass on the qualifications of the bondsmen, and might reject one or

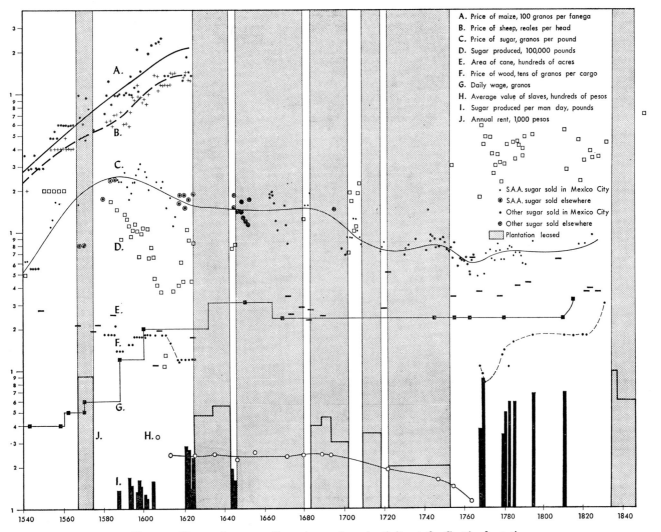

Legend (top right of figure):

A. Price of maize, 100 granos per fanega
B. Price of sheep, reales per head
C. Price of sugar, granos per pound
D. Sugar produced, 100,000 pounds
E. Area of cane, hundreds of acres
F. Price of wood, tens of granos per cargo
G. Daily wage, granos
H. Average value of slaves, hundreds of pesos
I. Sugar produced per man day, pounds
J. Annual rent, 1,000 pesos

• S.A.A. sugar sold in Mexico City
⊙ S.A.A. sugar sold elsewhere
• Other sugar sold in Mexico City
⊙ Other sugar sold elsewhere
▢ Plantation leased

Figure 3. Changes in commodity prices and productivity at the Cortés plantation

more of them, requiring the bidder to find substitutes or lose the opportunity to lease the plantation. When agreement had been reached on this point, the Governor and the renter signed the contract.

A date was set for transfer of the plantation, and the step was considered so important that very often the Governor himself was in attendance; the lessee nearly always appeared. Each party appointed an assessor to represent him in the long process — sometimes a week or more — of taking inventory. The process was further complicated, beginning in the seventeenth century, by the custom of evaluating not only the condition but also the cost of replacement of each item. Numerous people might be present: one mid-eighteenth–century inventory (L341 E5) was attended by the Governor, the controller, the lessee, three assessors, two master smiths, three master carpenters, two master bricklayers, the

mayordomo of the plantation, and numerous unskilled assistants. Such an inventory might cost more than $500.

I have chosen to describe as typical conditions of leases the most important ones contained in the contract between Andrés Arias Tenorio and the Estate covering the period 1625–34. This is the third such contract made, the two previous ones dating from 1566 and 1575.

The first section of the Arias Tenorio contract contains a statement concerning the length of the lease and the annual rent to be paid: it was to run for nine years, at $4,666-5-4 annually. The first condition states that in the rental period all the rights in force while it was administered by the Estate should remain in effect; the same equipment, slaves, and land should be transferred to the renter; and the transfer must take place within twenty days after signing of the contract. The second condition, related to the first, guaranteed to the renter that the Estate would deliver the Indian fieldworkers assigned to labor at Tlaltenango; this extremely important clause was duplicated in most of the rental agreements, and could be enforced by the Estate because of its administration of local justice (giving it relatively certain access to labor). Arias Tenorio, on his part, agreed to pay the customary wages to the Indian workers, and the Estate guaranteed delivery of part of the local maize tribute customarily assigned to the plantation. In the case of the maize, the price was not specified, but the dependability of the supply was an important advantage.

The condition that usually caused the most difficulty concerned which party was to bear the costs of repairs and damages that hail, disease, frost, or other chance occurrences might inflict. In the Arias Tenorio contract, all repairs were to be made at the expense of the renter, and all risks involving slaves, livestock, and cane were also his. This last stipulation, particularly in reference to the damage done by frost, is markedly different from the first rental agreement, wherein the Marqués bore responsibility for such damage up to the amount of $9,000. It also differs greatly from later rental agreements, which usually state that a certain amount of the annual rental would be credited towards improvements; in these cases the annual allowance was summed over the rental period, and the difference between this sum and the valuations of the improvements went to the appropriate party.

Evaluation of the cane was extremely important because it usually constituted the most valuable item in the inventory. The Arias Tenorio contract provided for dealing with an excess of cane at the end of the lease, the Marqués promising to mill it and pay for it, or else allowing Arias Tenorio to mill it in his own neighboring mill without payment to the Marqués for the use of the land on which it was growing; however, a limit of 60 machos of cane was set as allowable excess, the remainder going to the Marqués.

The renter agreed to provide religious services, clothing, and medical care for the slaves, usually the second most valuable item in the inventories. The deaths of any old slaves described as "ynutiles" by the assessors and any under six years of age at the time of conveyance were not to be the responsibility of the lessee, but in the case of death of others the renter had either to replace them or to give cash compensation. All those born

during the lease were to be the property of Arias Tenorio, who was also responsible if females died in childbirth. This stipulation differed from its counterparts in later contracts in that usually in the latter the Marqués and the lessee each received half of the slaves born. In practice, the Marqués usually bought all the slaves born under these conditions so that families might remain intact.

Arias Tenorio bore responsibility for all taxes, tithes, alcabalas, costs for all crimes committed by slaves, damage caused by livestock, and payment of the salaries and other costs of doctors, priests, and other officials of the plantation.

The parties also agreed to protect each other's right in the plantation, with the renter not allowed to sublet it without permission from the Marqués, nor the latter allowed to rent or sell any part of the land or equipment. A clause omitted from most other contracts stated that Arias Tenorio might rent the sugar store of the Estate in Mexico City. Finally, both parties testified that they had entered the agreement in good faith and pledged their belongings to its fulfillment; Arias Tenorio pledged his mill Amanalco, promising not to sell it during the lease. The parties also agreed that costs of debt collection were to be borne by the lessee, and that both would submit disagreements to the local justice. The contract was witnessed by three persons, all vecinos of Mexico City, and signed by the parties before the royal notary Alonso Hidalgo Santillán.

Characteristics of Lessees. Appendix B (p. 110) describes the major characteristics of lessees of the plantation. Nearly all were vecinos of Mexico City or of Cuernavaca, and all may be assumed to have been familiar with the condition of the mill. Either they had been engaged in the leasing, ownership, or management of other plantations in Morelos (the most numerous group), or else they were merchants in Mexico City who sold sugar there or began to do so immediately after receiving the plantation. Those who were plantation owners arranged for sale, probably exclusive, through an already established merchant—in some cases a relative—in Mexico City, and the merchant generally both supplied cash to meet weekly payrolls and other costs and supplies, and expedited affairs in the capital. This kind of relationship seems to have been standard in New Spain nearly from the beginning of the industry.

Many of the lessees held official positions in Mexico City, or in Cuernavaca, at some time in their careers. Their choice of bondsmen indicates in most cases very strong connections with the merchant class; very few of the bondsmen after the sixteenth century were themselves planters, although they may have been associated with plantations as sugar vendors in Mexico City. These merchants, with close ties — of blood, marriage, business, and especially debts — to the planting community, obviously deserve considerable study.

Profitability of Rental. It is difficult to determine whether the plantation was more profitable when leased or when administered directly by the Estate because so few lawsuits contain final decisions of the courts, and there is also little information concerning the collection of amounts claimed by the Estate. Because of the availability of data, I shall limit discussion to the period 1700–60, but it is worthwhile to note that the second lessee of the sixteenth century was claimed by the Governor to owe $36,000 (L282 E2), a very large sum.

In 1760, after several decades of low sugar prices, the Governor called for a review of the record of the hacienda since 1700, in order to persuade the officials of the Estate in Spain that it should be sold. The summary in Table 6 (p. 128) is based on this perhaps biased review (L341 E14). The controller calculated that the total loss to the Estate was $76,214, more than the total shown in Table 6; the difference may be due to the fact that some claims were still before the courts. Although the period when the mill was in administration from 1718 to 1721 was omitted by the controller from his summary, the fact remains that it had been losing about $1,000 annually for nearly sixty years. After the rental ended in 1763, it was administered by the Estate for about seventy years, and seemed to do better during that time.

Administration of the Plantation by the Estate

This section has a double focus, dealing on the one hand with a major argument against Estate administration — that it was difficult to get cash to finance operations — and, on the other, with the favorable argument that the Marqués' sugar store in Mexico City was the source of much of the cash. The rest of the cash came as part of a complex flow of small change through the mercantile network in Cuernavaca and Mexico City.

Coinage was frequently in short supply in colonial Mexico, especially from about March through August or September, when large amounts of silver were being collected for the flota destined for Spain, as well as in some of the months following its departure when stocks of coins were only beginning to accumulate. In spite of these chronic and seasonal deficiencies, the workers had to be paid, at the rate of about $500 weekly in the nineteenth century, and in coins of small denominations: Fanny Calderón de la Barca (p. 399) briefly describes seeing in 1841 a distribution of a week's pay at one of the haciendas of Puebla, where there were "mountains of copper piled up on tables in front of the house." Once received by the workmen, the coins continued on their paths to the merchants, the plantation stores, tax collectors, and churches, but these local sources were on numerous occasions inadequate to meet local demand, and the deficiency could be met only by appeal to the Estate in Mexico City or, in the case of managers of other plantations, to some other source in the capital. The difficulty of obtaining money to finance the day-to-day operations of the plantation, together with the generally low price of sugar through the seventeenth and the first half of the eighteenth century, made rental of the plantation seem the more attractive of the two alternatives available to the Estate at this time, sale being out of the question.

Independence did not make obtaining cash any easier. After Alamán found himself forced to assume the administration for want of a renter in 1847, he decided (*ibid.*, p. 457) that it would be necessary to open the hacienda store in order not only to satisfy the needs of the workers for clothing and food but also to obtain cash as well as decrease the need for it. The decision to re-open the store came after he found himself behind in paying wages at the hacienda, a circumstance that required him to borrow on the Hospital account at an interest rate of $1\frac{1}{2}$ per cent per month (*ibid.*, pp. 438; 357–58). These difficulties had been complicated earlier, in the mid-1830's, by the fact that copper coins were practically the only kind available, and their value fluctuated widely within short periods of time.

Libranzas. Much of the cash was received through issuance of libranzas, or drafts, by the plantation administrator on the Governor. Few of the thousands administrators must have issued have survived; most of the survivors date from the 1750's (L341 E4). The following example from that group serves to indicate their nature:

[To] Sr. Don Joseph de Asso y Otal
Por esta se servira Vˢ. de mandar pagar la cantidad de duszientos pˢ. a la Boluntad del sʳ. Dⁿ. Bentura de Salazar bezino y mercader de la Billa de Cuernavaca Por otros tantos que tengo aperzevidos de dicho sʳ. **Para**

avios desta Azienda que por esta y su contenta seran bien dados

Sⁿ. Antt^o. Tlacom^{co}. y Julio, 6 de 1754

Augustin Fer^z. [Administrator of the Mill]

Son 200 p

Paguese esta libranza de los productos de el azucar q estan a cargo de el encomendero dⁿ. Juⁿ. Antonio Parrazar; tomando razon de ella en la Cont^{oria}. Gral de el estado.

Mex^{co}. y Agosto 21 de 1754

Joseph de Asso y Otal

Mexico Ag^{to} 31 de 1754. Tome razon

Feliz Lince [Controller of the Estate]

All of the above occupied one side of a sheet of paper of approximately legal size; on the reverse appears only the signature of Bentura de Salazar. Others might add on the reverse a sentence to the effect that they had received the amount stated in the order. On one (*ibid.*) the seller of the Estate sugar wrote: "No pago esta Libranza por no tener vendida la Azucar la que pagare luego que se complete. . . ."

The sequence of events involved in issuing an order to pay appears clear from the example. To pay for supplies or labor, the administrator of the plantation wrote an order to pay against the governor, who approved it on receipt; the controller then received the libranza, usually on the same day, and it was usually presented for payment within the same building where the sugar was sold. Finally, it was filed by the sugar seller with the controller when he presented his accounts for approval.

Libranzas were accepted in lieu of cash for goods as well — the purchase of, for example, a hundred cattle might be accomplished with a libranza; but in general the purpose of libranzas was to obtain from local sources enough cash to pay the workers. The records do not show that any charge was made for this service, and from the point of view of those who advanced the money the advantages appear obvious when contrasted with the alternative of shipping money from Cuernavaca to Mexico City to settle their own accounts, particularly since the credit of the Estate was always good. It may be assumed that they, in turn, received the money that they exchanged for libranzas from local persons with whom they did cash retail business. The advantages of the libranza are suggested by the history (L341 E4, 14 May 1754) of a libranza for $4,000 made by the plantation administrator to the same Bentura de Salazar, which went by successive endorsements to

Bernardo Campi Delgado, Joseph Vicente de Ojeda, and finally to Gaspar de la Concha (resident of Mexico City) on account of Joseph de la Torre Calderón, resident of Vera Cruz. The exchange was rapid, inasmuch as the original was dated 8 May, and the libranza was presented for payment to the Governor on 13 or 14 May, less than a week later. It seems likely that the endorsers were merchants. Most libranzas did not pass through so many hands as did this very large one, but most seem to have been presented for payment within two weeks.

The mayordomos were expected not to issue drafts so freely that payment would be difficult; instead, a rough balance between the value of drafts and the value of shipments of sugar to Mexico City (minus costs of shipment) was supposed to be maintained. In 1833, for example, the Apoderado General Mariano Tamariz wrote to the administrator of Atlacomulco, Mariano Rubio, as follows (L348 E61): "Le reiteri a V el encargo de que no libre si no es con ocho dias a lo menos de vista y solo por la cantidad que importen las remesas de azucar que se hayan recibido aqui, teniendo presentes los gastos de alcabalas, fletes, mermas y cargadores."

The widespread use of drafts in conducting business in Mexico is shown by the fact that in the 1840's Alamán was buying libranzas from the government on behalf of the Estate. The government received these drafts from merchants for payment of taxes and duties ("derechos") and discounted them at the rate of 1 per cent per month; Alamán felt that these purchases were safe: "como las pagan los dichos comerciantes, sin tener nada que ver ya con el govierno, no hay en ello riesgo ninguno" (*Documentos*, Vol. 4, p. 510).

Money from Sale of Sugar in Mexico City

The Estate maintained a sugar store in Mexico City on the lower floor of the palace of the Marqués. A principal function of the mayordomo of this store was to accept the libranzas of the mayordomo of the plantation, paying cash to those who submitted them. By contrast, at least in the seventeenth century, the value of goods sent by the mayordomo of the store to the plantation was not large, running between $1,000 and $1,500 annually (L263 E2), whereas the payments on presentation of libranzas might be much larger than that, as they were particularly in the eighteenth century (Table 7, p. 128). For example, in approximately four years (1770–74; L231 E5) the mayordomo of the store in Mexico City, Jose Ramón Mateos, redeemed about $45,500 worth of libranzas from the mayordomo of Atlacomulco, a rate of about $11,000 per year, but he sent much less goods, by value, to the plantation.

Management of the store in Mexico City followed a pattern by now familiar. Much of the time it was rented when the plantation was, and the rest of the time a mayordomo was appointed and his accounts treated in the same way as those of the Governor and the mayordomo of the plantation. Similarly, he was required to have bondsmen, was paid a salary, and in many cases a commission on sales was given to him. Table 8 (p. 128) shows the range of his salary over time, including an item for food; much of the food went to workers (carreros) in the store. From 1702 to 1825 the salary, including the food allowance, was $500 or $600, with one especially valued administrator receiving $1,000 in 1779. The governor raised the salary of the administrator Augustín de la Cassa in 1709 by $100, to $500 per year, justifying his action in the following words: "a fin de que pudiese mas commodamente mentenerse, y en consideracion de ser premio moderado, respecto de lo laborioso y pesado de su intendencia, que desempeno a toda satisfaccion mia, y con ventajosa utilidad del negocio, sobre que puso continua, y extraordinaria diligencia, meritos todos persuasivos . . . conque aun no queda competentemente satisfecho en mi inteligencia" (L30 V58 E1).

In the 1770's, in addition to a salary, a commission of 4 per cent of the total value of sales was given to the mayordomo. There was, further, an upper limit set on wastage and weighing errors, 3 per cent during much of the history of the mill, which represented a potential profit to the mayordomo if he could keep losses below that proportion of the sugar received from the mill. The interests of the two mayordomos of mill and store were here at variance, since the potential loss depended in part on the quality of sugar made at the mill. In the 1580's, for example (L208 E7), a complaint was lodged by the mayordomo of the store, Carlos Pérez, concerning very high losses from wet sugar and poor sugar that was not saleable, in which he included an affidavit from another sugar seller in Mexico City stating that losses of 10 per cent might be expected on sugar as poor as that Pérez had been receiving. Differences among scales amounted commonly to differences of 2 per cent in weights recorded, according to another witness, and all stated that handling the sugar, particularly sugar that crumbled easily, would result in losses as well. An allowable loss of 3 per cent does not seem excessive under these circumstances.

Sales were not made exclusively through the Estate store. In 1680–81 (L257 E15) the mayordomo, Diego Pinsón, vecino and mercador of Mexico City, sold Atlacomulco sugar in his own store as well as in the Estate store. In the late sixteenth century, large amounts of sugar were sold at public auction at the same time as the Estate maintained its store (L240 E1B) : one such lot of 1588 involved 2,000 arrobas of white sugar sold for $12,000. Sales of this size undoubtedly were made to confectioners and were often regarded as unfavorable to the Estate, since the price of such large lots was generally lower than that of smaller quantities and it depressed the market price as well. The practice of offering lower prices on large quantities was nevertheless standard in the late eighteenth century (L232 E1 D3). In the nineteenth century, it may have been common for plantation owners to receive their necessary advances from merchants on condition that the latter be exclusive sellers of their sugar. In December of 1847 (Alamán, *Documentos*, Vol. 4, p. 460) one Martínez del Campo of Mexico City offered to take all Atlacomulco sugar of that zafra in exchange for advances, and Alamán was relieved that the arrangement was made, since for awhile at least it removed the nagging necessity of finding the cash needed to finance operations at the plantation.

The Estate store paid rent to the Hospital de Jesús. The amount was $60 in the 1590's, double that figure in the early eighteenth century, and $294 through the late eighteenth and early nineteenth centuries. The accounts of the mayordomos include as major items only their own salary, the rent, and miscellaneous items called gastos menores that seldom exceeded $150 in the eighteenth century. Probably for much of the colonial period a figure of $1,000 represented fairly well the annual expenses of the store.

The mayordomo or administrator of the store also paid the freight charges on the sugar and the alcabala, which always amounted to much more than the cost of maintaining the store.

Cash from the Hacienda Store and Sale of Molasses. The principal items sold at the hacienda were yard goods and molasses, of which the latter was expected to bring in the most money and apparently long antedates the other as a source of cash. From the beginning molasses was sold locally, and information concerning its production and sale at the plantation is included below.

The hacienda store probably was first established in the eighteenth century, or in the late seventeenth, but its full development did not occur until the last half of the eighteenth century. Its principal functions may have been to accommodate both the needs for food and clothing of the growing mestizo and mulatto population not associated with Indian pueblos and also to bind them through debt to the hacienda, at the same time as it brought in cash needed to operate the plantation. The last function seems to have become much more important after Independence, and at the same time reliance on pueblos for field labor was considerably lessened since the Indians were no longer obliged to contribute labor to the haciendas. Thus, after Alamán found himself forced in 1847 to assume the administration of Atlacomulco for want of a renter, he decided (*ibid.*, p. 457) to re-open the store; in 1848 he expected profits of $2,000 (*ibid.*, p. 476) from this operation.

Another expedient widely resorted to after Independence, and equally widely condemned as a vicious feature of the plantation system, was that of manufacturing and selling aguardiente at the haciendas. As with the stock of the store, the product was destined for sale to the laboring classes, and the major justification for the practice had to be that it supplied cash for financing weekly operations.

Cash from the Alcaldes of Cuernavaca. In the early eighteenth century, possibly also in the late seventeenth, the administrators of the mill had to rely on receipts of cash in small amounts from the alcaldes and alguaciles of Cuernavaca, who had collected the money as tribute payments; thus, rather than sending all the money to the Governor in Mexico City, they turned some over to the administrators of the plantation. From time to time the Indian governors of the pueblos paid at least part of their cash tributes to the plantation directly. As Table 7 shows, this form of transfer fell into disuse sometime after 1720. In the period between 11 March 1702 and 9 February 1703, the receipts from the alcaldes consisted of thirty-one payments, mostly of $100 or $150, the total being $5,020; this mode of supplying money was in use when numerous plantations had been abandoned in the region because of the low price of sugar and the inability to obtain cash to run them (L30 V56 E1).

Cash from the Governors. The Governors were finally responsible for supplying cash, but they did not part with it freely, nor did advances arrive from them without solicitation. The fragments of correspondence between Governors and mayordomos that survive suggest rather that the latter had to request cash, and most of their correspondence contains this request, together with information and requests for supplies. The sending of specie in small denominations was usually accomplished not by special messenger, but more often by the return trip of the mule trains that had carried sugar to the capital.

Table 7 summarizes information from 1718 to 1828 concerning the principal sources of cash for the plantation. It shows clearly the dominance of the libranza, whose total values for any year should be compared with the total expenses of running the plantation; the total expenses are taken from the data totals of the annual accounts, so they are a close approximation to actual expenses. The amounts obtained by issuing drafts had a greater range than did total expenses, but the last three years in the list suggest that an amount in the neighborhood of $20,000 was obtained by drafts; probably this pattern had been established by 1785 ($23,-386), since the very low figure of 1811 expresses the difficulties of obtaining cash while the Revolution was running its course. Cash from tributes, often obtained by issuing libranzas to local officials, seems not to have been so important as cash obtained from local merchants. After Independence, the sales of molasses and sugar at the hacienda became much more important than they had been previously, because selling molasses to make alcoholic beverages was no longer prohibited.

IV

The Acquisition of Land and Water

THE major feature of the location of the Cortés plantation was its position within the largest subtropical basin near Mexico City, combining a climate suitable for sugar cane and other subtropical crops with nearness to the major Mexican market for sugar. Of great importance also was the proximity of the steep and high escarpment of the volcanic plateau. The volcanic materials issuing from vents near the rim of the plateau formed a complex mosaic of mudflows, lava flows, and ash showers, all subsequently dissected, on the basin floor that offers a wide range of agricultural opportunities. The escarpment, reaching more than 10,000 feet above sea level, experiences a heavier precipitation than occurs at lower altitudes; much of the precipitation that falls in the cool uplands emerges at the base of the escarpment as springs, and some of the rest moves into the basin as streamflow. This combination of physical features and proximity to Mexico City, together with the dense and large aboriginal population, must have seemed very attractive to the Spanish settlers.

Within this region, the sites of the Cortés mill had a number of advantages and one major disadvantage. The plantation was the nearest of all the plantations of Morelos to Cuernavaca, site of the Cortés palace, and the journey between them could not have required more than half an hour by horse. In addition, time and costs of travel to Mexico City were lower than for any other plantation in Morelos. The plantation was also nearer the major spring line than most other mills, and downstream from no other major user, important where irrigation was indispensable to cane culture. In spite of these advantages, however, both sites suffered from the frost hazard associated with relatively high altitude (1,400 and 1,700 meters), particularly Tlaltenango.

Landforms and Geology

The state of Morelos has a subrectilinear shape roughly outlined on three sides by mountains. To the north is the prominent escarpment that rises to a series of volcanic peaks, culminating at the northeast corner of the state in the famous cone Popocatepetl; a less well–defined set of ranges lies along the west side, and another range is more or less coextensive with the state boundary on the south. To the east, the boundary with the neighboring state of Puebla is not marked by conspicuous physical features. Within the boundaries of Morelos, the most important physical feature is a highly dissected accordant surface consisting mainly of volcanic debris, and divided into two parts by several low, meridional ranges that separate the Cuernavaca and Cuautla Amilpas basins.

Two major formations (Fries, 1960) underlie the extensive dissected plains. In the northern part of the basin, the Chichinautzin group of Quaternary basaltic and andesitic lavas predominates, extending from the escarpment slope onto the gentler slopes of the basin; in one case, a long narrow tongue of this material reaches well past the eastern side of Cuernavaca almost to Zacatepec, near the southern edge of the basin. Most of the rest of the more or less plane surface is underlain by the Cuernavaca formation, consisting of a wide variety of materials of which the most conspicuous and probably the most extensive is a conglomerate with large boulders in a matrix of volcanic ash.

The Chichinautzin group and the Cuernavaca formation differ in their surface expressions. The most striking characteristic of the Chichinautzin group is aptly described locally by the words malpaís or pedregal. In some places it is extremely rocky and difficult to traverse, with little soil visible, yet covered by fairly dense xerophytic brush and cacti. In other places where the rock surface is less obviously continuous, numerous small rocks and boulders nevertheless abound on the surface and the soils are very thin. The Cuernavaca formation, by contrast, has a much less broken surface, and its soils in general were probably originally somewhat deeper than those overlying the Chichinautzin group, although at the present time they are in few places more than about six inches deep. The grass cover of the Cuernavaca formation is more continuous than that of the Chichinautzin group, and the soils serve better for temporal in general than do soils of the latter.

The two formations differ as well in their hydrographic characteristics. The softer, poorly consolidated Cuernavaca formation has experienced much greater dissection, locally approaching a badland condition, and the drainage courses are much more evenly distributed than are those of the Chichinautzin group, where deep, steep-sided, and in many cases impressive barrancas have developed. There seems to be a much greater number of springs in the latter group, with the result that the major permanent streams rise in and pass through it, whereas the parts of stream courses that occur in the Cuernavaca formation are more likely to be ephemeral.

Of more importance for sugar cane cultivation than the Cuernavaca and Chichinautzin materials are the discontinuous and relatively small deposits of Quaternary alluvium that occur along the major streams, and the mixed clastics and alluvium of the same age that extend from Yautepec and Xochitlán to Ayala, south of Cuautla. In this gently sloping area between Yautepec and Ayala is the most extensive area of cane in Morelos; the valley of Jojutla, the second most extensive area of the same materials, is by contrast much smaller. Although irrigation is easiest on these alluvial surfaces, cane cultivation is by no means restricted to them, extending indeed onto extremely rocky Chichinautzin surfaces and over a large area of the Cuernavaca formation that lies north of Lake Tequesquitengo.

Clearly, however, these alluvial materials and clastics, with their deeper soils, have been favored over others for cane cultivation since the sixteenth century, and more of their surface is still occupied by cane than by any other crop.

Atlacomulco depended on cane grown on a relatively small nucleus of alluvium, but some was also grown on adjacent relatively flat areas of the Chichinautzin group. When the mill was at Tlaltenango, north of Cuernavaca, much of the cane came from adjacent areas of the Cuernavaca formation; these soils did not stand up well to cultivation, and after only about fifteen years of use were no longer productive (L282 E4; L285). The raising of livestock, an extremely important part of plantation operations, depended in large part on pasture and forage growing on a large area of Chichinautzin materials north of the canefields, as well as on a small area of the Cuernavaca formation west of the main part of the plantation on the adjacent Rancho Guimac, acquired by the Estate for this purpose in the late eighteenth century.

The Mill Sites

Tlaltenango. Ruiz de Velasco (1937, pp. 135–37) discussed the evidence remaining in this century of the site of Tlaltenango, identifying it by means of some remnants of masonry that resembled a support for a waterwheel; this site is directly adjacent to the Cuernavaca–Mexico City highway near the circle in whose center stands a likeness of Emiliano Zapata. Quoting a local resident, Ruiz de Velasco stated that parts of the walls were still standing in this century, and contained some identifying marks as well as the date 1535. The distance to the Cortés palace in the central part of Cuernavaca is about five kilometers.

Physically, the site lacks advantages (Fig. 4). The land drops almost immediately to the west to a deep barranca, whose waters are fed by springs near Santa María Istayuca, and several hundred feet east of the site is another deep barranca. Several barrancas, 50 to 100 feet deep and with nearly vertical sides, are distributed in roughly parallel fashion in the dissected area around Tlaltenango. In some places, the ridges are barely wide enough for a road. It is necessary to cross at some point a deep barranca and travel several miles from the site before reaching any extensive area of land suitable for cane, except downslope toward the south and through Cuernavaca.

The scarcity of level land near Tlaltenango made it

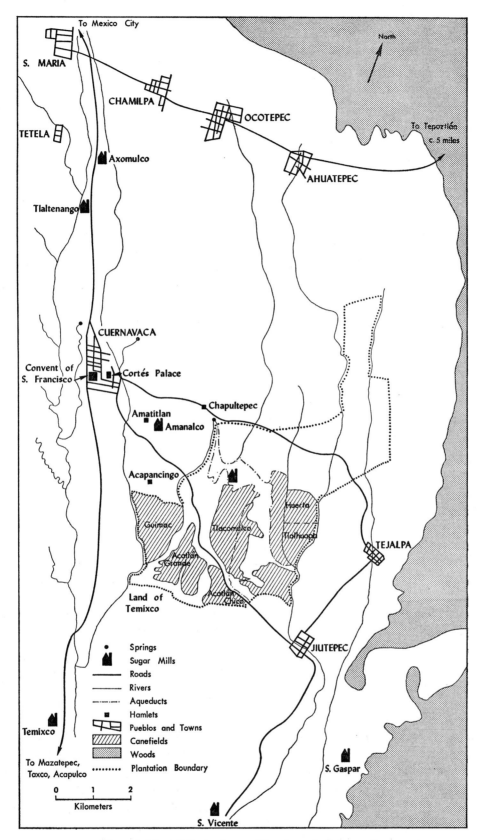

Figure 4. Map of the environs of San Antonio Atlacomulco

27

impossible to irrigate much cane there. The most feasible intake sites for an irrigation system were about four kilometers north, at the springs of Istayuca, the only nearby source with a volume of flow sufficient for both irrigation and milling. The water was thus transported some distance to the mill, irrigating canefields on the way, as well as some south of the mill, and supplying the city also. The resulting pattern of field distribution on the ridge was narrow and probably about ten kilometers long.

Axomulco. It is not possible to fix with certainty the site of Axomulco, but L282 E4 contains enough information to justify its placement as shown on Figure 4. It was north of Tlaltenango, and possibly on the road from Cuernavaca to Mexico City. Much of the land was supposed to be within or contiguous with the pueblo of Tetela — probably east of the town, since the barranca that passes the northeast corner of the town is very deep and the Cuernavaca formation, less suitable for cane, lies west of the town. The water came from the springs of Istayuca, as did the water of Tlatenango (L282 E4).

San Antonio de Atlacomulco. Concerning the site of Atlacomulco there is no doubt, since the ruins remain. This mill was sited in a way common to many of the mills of Morelos, strongly related to the flow of water and the juxtaposition of contrasting types of terrain. Here, as in many other places, the break between alluvium and other materials is very sharp, being marked by a strong contrast in slope and altitude. The alluvium may lie as much as 200 feet below the surface of adjacent Chichinautzin or Cuernavaca materials, providing good potential water power. Atlacomulco was built at the base of what may be a flow of volcanic debris, about 40 feet below its upper surface. The water was led to the mill by aqueducts and ditches from Chapultepec Springs about 1½ kilometers to the north, then conducted by a conduit beyond the edge of the cliff to fall onto the wheel.

Acquisition of Land

Of the two major modes of land tenure, rental and ownership through purchase, the first was extremely important in the sixteenth and early seventeenth centuries, but because of the purchase of the tract of approximately 420 acres called Tlacomulco in the 1620's the second mode of tenure became dominant. No change of importance occurred after this time until independence was gained from Spain, when ownership became regarded as the most desirable form of tenure; after 1851 no land was rented for cane production.

RENTAL OF LAND

Most leases ran for nine years, which seems to have been the maximum length, and renewal was usually easy. Many sixteenth- and seventeenth-century leases were for periods of six years, and a large number of relatively small pieces of land were rented for periods of two to six years in the same period. A typical lease is reproduced in Appendix C (p. 113); it dates from 1597, and the parties were the Estate and the neighboring pueblo of Jiutepec (L263 E14).

The lease with Jiutepec was registered before an official of the Marquesado in lieu of a royal official. Characteristically such documents include justification for the rental on the part of the Indians, together with a statement that they have retained sufficient land for their own crops, but such a statement is missing from this example. The clause stating that the owner cannot reclaim the land before termination of the period specified in the lease without providing land of equal amount and quality is also typically included, as is the statement that the renter has to pay the total rent whether or not he continues occupying the land for the entire leasing period.

The earliest complete information concerning the amount of land rented by the plantation describes conditions in 1549 (L285), and the information is summarized in Table 9 (p. 128). The annual rental rate applicable to all the pieces was $1 per 1,500 square brazos (approximately one acre). In the light of this information, it appears likely that the pieces of land in the category "not identified," whose annual rental price was $70, was probably nearer 70 acres than 120. Perhaps the total to continue in rental was thus about 145 acres rather than 204, making a grand total of approximately 425 acres of rented land in 1549. Since, according to the inventory of 1549, about 531 acres were in cane in that year, it appears that about 80 per cent of the land used for cane was rented.

Although the outcome of the lawsuit brought by the Indians in 1549, the source of this information, was unfavorable to the Estate, with the Audiencia decreeing that the latter must pay compensation and return most of the land to the communities, there was little change in dependence on rented land for the rest of the sixteenth century. Thus, in the following decade, in a Relación of 1556 (L267 E26) describing the landholdings of the Marqués, it was mentioned that cane land was available on six-year leases, and that the entire rent had been paid in advance on many of the plots. It is not possible to calculate the proportion that rented cane land bore to the total land in cane, but later statements, such as a listing of 1590 discussed below, suggest that it must have remained high. Even the site of the mill was rented in 1549 and 1556, at the extremely low price of $12 per year.

When the mill returned to Estate administration in 1581 there began a series of mayordomo accounts with data sufficient to describe the characteristics of leasing until 1624, when the plantation itself was again rented. In 1581 twenty-two rented pieces were in cane, with the highest individual rental fee only $8 per year, paid to Don Toribio de San Martín Cortés for a piece whose dimensions were not given. In 1586 twenty-seven Indians received $357 for the rent of canefields, and in 1590 one-third of all the land in cane was rented, of which half belonged to Don Toribio. Of land owned by the Marqués, approximately 200 acres were in Atlacomulco and only 50 in Axomulco.

Table 10 (p. 129) lists all the information available, principally from accounts of the plantation, concerning payments of rent for land. The total paid annually for rent varied little in the years 1581–99 and the final year, 1831, being about $200 at both times. What is most important is the fact that whereas before 1624 many small payments were made to a large number of individuals, in the last part of the century the number of such payments seems to have declined. In most of the seventeenth, and all of the eighteenth and early nineteenth centuries, payments were made to the Indian governors and principales of pueblos and towns for relatively large blocks of land rather than for small individually owned fields, and the purchase of these contiguous blocks gave final form to the plantation.

PURCHASE OF LAND

The purchase of these large tracts will be discussed according to the town or pueblo that owned them. Figure 4 shows the necessarily generalized positions of the tracts within the boundaries of Atlacomulco. Although these existed as named and bounded regions for the purposes of rental and purchase, available maps show only in a general way the positions of some of them, and no map except one dated 1760 shows shapes (but it is probably inaccurate). Nearly all the names of these tracts are Indian, so it is likely they were as-

signed their names in pre-Conquest times. It is not clear why they should have been named, hence implied to have some sort of unity; the latter is not obviously physical in the sense that they are bounded by physical features such as rivers or cliffs, because there are few major physical features on this relatively flat plain. It will be seen that lack of information about their shapes contributes considerable vagueness to the discussion.

The blocks to be discussed are Tlacomulco, the largest, which gave its name to the plantation and was its core from the mid-sixteenth century; Guimac, a ranch acquired by the Estate in 1791 for pasture; the two Acatlanes, Grande and Chico, at first consisting of irrigated pasture and later converted to canefields; La Huerta, a relatively small tract of about 55 acres, an irrigated orchard until the mid-eighteenth century, when it was changed to cane; Tlalhuapa, a tract of pasture until about the same time. The last four were all purchased by Lucas Alamán in the mid-nineteenth century.

Other tracts named in seventeenth and early eighteenth century accounts were called Tesontepec, Acecentla (at one league or a "larga distancia" from the mill (L269 E2)), and Zacualpa. The latter two were first mentioned in the inventory of 1606 and apparently belonged to Jiutepec; I have no information on the final disposition of these three.

These named tracts apparently encompassed all of the irrigated cane land of the plantation, with the rest supporting only rather poor pasture on thin soils. The pasture land seems not to have been divided into named tracts as was the irrigated land.

Cuernavaca Land. The core of the plantation was a large area of fields called Tlacomulco (or Talcomulco, Tlalcomulco, and finally Atlacomulco), first mentioned in a lawsuit of 1549 (L285) between the Marqués and the Indians of Cuernavaca. Information in this document makes it clear that a fairly large area was in use by the plantation in Tlacomulco by the early 1540's, since an aqueduct had been built to the tract. At that time, large numbers of persons owned fields of varying size in Tlacomulco, a pattern that persisted until at least the early seventeenth century; in the account of 1606–7, for example, were noted rent payments to the Convent of San Francisco in Cuernavaca and to a private person for fields there. The Marqués also owned land there, because another part of the tract was described in a document of 1602 (L263 E14) as "tierras que entiende que son del vinculo del Marques que andan con el ingenio," and more briefly in the same document as "tierras del mayorazgo"; still other fields in Tlacomulco were described as rented. The inventory of 1611 lists fields used for cane, both rented and owned by the Marqués, in the "pueblo" of Talcomulco.

By far the largest part of Tlacomulco was the tract of four caballerías, approximately 420 acres, described in a rent receipt of 8 May 1620, and apparently first rented from Don Toribio de San Martín Cortés, Indian governor of Cuernavaca in the late sixteenth century. The rent was paid as a censo perpetuo of $80 annually. Since this tract appears never to have been a source of controversy between Cuernavaca and the Estate, it is not surprising that there exists no contract for its rental or apparent sale to the Estate. It is possible that it was sold to the Estate about 1620, since according to a one-page receipt dated 26 April 1620, the testamentary executors of Don Toribio's estate, one of whom was then the governor of Cuernavaca and neither of whom understood Spanish, had received $225 from the mayordomo of Tlaltenango as the balance owing when the property was sold: "es el resto q se debia hasta q el dho pedazo de tierra vendieron. . . ." This important tract is thus a mysterious entity about which all important facts except its name, approximate size, and rental value are unknown. There is nowhere any description of its form or location, in spite of its importance and frequent mention in documents of the late sixteenth and early seventeenth centuries, and it is not mentioned after the translocation of the mill, apparently to the tract, in 1642. However, the land was cheap, and for that reason we may assume that water rights were not associated with it.

Three other pieces of land were rented from Cuernavaca. Two, adjacent to each other, were called Acatlán Grande and Acatlán Chico; the other was called La Huerta or El Potrero de la Huerta. In 1814 the mayordomo Calleja wrote that there existed no titles or other documentary proof of ownership of these lands, but that the Indians of Cuernavaca were nevertheless the owners (L278 V48 E11), and I have found no other proof of ownership.

Calleja also mentioned that the Indians had been receiving rent for the Acatlanes without protest since before 1779, to the amount of $150 per year, so the land did not become a source of trouble until after the Independence movement began. Thus in 1814, after expiration of a lease agreement (of which no example exists),

the Ayuntamiento of Cuernavaca decided that Atlacomulco must give up the Acatlanes because they belonged to the Indians. Atlacomulco must have been the highest bidder for these tracts, if they were submitted to public auction at the time, since it does not appear that the plantation lost the use of them. By 1827 (L27 V48 E11) the mayordomo wrote that it seemed likely that the rent of the Acatlanes and La Huerta would be raised one-quarter or one-third by Cuernavaca, implying that these tracts were not disposed of by public bidding, but there is no record that the price was raised. Circumstances were very different in 1849 when, at the expiration of the lease, the Acatlanes and La Huerta were auctioned and rented to the pueblo of Acapanzingo, adjacent to the western side of the plantation, at an annual rent of $352, the contract to run for five years. Besides the plantation and the pueblo, a private party, one Ignacio Betancourt, was involved in the bidding. The importance of these lands to the plantation moved the Estate to offer to exchange the use of its potrero or rancho de Guimac, adjacent to the southern boundary of the pueblo, for the use of the Acatlanes and La Huerta. The pueblo consented to the exchange, but it is surprising that the Estate was willing to risk losing through public bidding lands that it had been using for many years without prior assurance that a satisfactory exchange could be effected (L417 E14).

La Huerta was apparently not a large tract, but its size cannot be stated with certainty. According to the 1760 map, it consisted of one-half caballería and one solar (about 55 acres) and approximately the same dimensions were assigned to it in 1743 (L27 V48 E9 D1). It lacked a grant of water (*ibid.*), so the rent was fairly low. Until the mid-eighteenth century its major feature was a small orchard, not identified except as "arboles frutales muy util," that were cut down by Tomás de Avila Romero about 1743 in order to plant cane in their place (*ibid.*). It was first rented by the Estate in the late sixteenth century, and is clearly identified in some accounts after 1600 by its rental price of $30; it had belonged to Don Toribio de San Martín Cortés, and the rent was used to pay for an annual mass for his soul (L27 V48 E9 D1). The latest information concerning payment of rent for La Huerta alone appears in the account of 1822, when it was still $30; after 1831 the rent may have risen, since $200 was paid for both La Huerta and the Acatlanes, which together previously had cost only $180. La Huerta was not always planted with cane; in 1718–21, for example, it was an agostadero, used for pasture in the dry season, and in 1822 was referred to as a potrero.

Lucas Alamán must have felt that the tenure of Atlacomulco in the lands rented from Cuernavaca was not very secure after Acapanzingo successfully bidded for rental of La Huerta and the Acatlanes, so through his efforts the Estate bought these tracts from Cuernavaca in 1850. All information on this transaction is in the short document L417 E12, a summary made of the original in 1873 at the request of the administrator of Atlacomulco. The sale had been approved by the Legislature of the State of Mexico on 30 November 1850, and the Estate agreed to pay $10,000 for the three pieces, half in cash and the rest in instalments that were completed in 1860 (*ibid.*).

Land of Jiutepec. About five kilometers southeast of the mill was the Indian pueblo of Jiutepec, whose lands and those of its tributary to the north, Tejalpa, were contiguous with land belonging to the plantation along its entire eastern side. Even before the mill was moved from Tlaltenango two pieces were rented from Jiutepec for cane, Acecentla and Tlalhualpan. The history of Acecentla is obscure. It may have been incorporated within the plantation in the seventeenth or eighteenth century, and it never was a source of controversy. Alternatively, Acecentla may have been near Zacualpa downriver from Jiutepec and never was incorporated within the plantation.

Tlalhuapan was first used as an adjunct to Atlacomulco about 1645 when one of the renters occupied it in an unused (eriasa) state (L27 V48 E9 D2). The first reference that I have found to its use by the plantation when it was administered by the Estate applies to 1707 (*ibid.*) and a statement describing conditions from 1718 to 1721 contains the information that $22-4-0 was paid to the Indians of Cuernavaca (*sic*) for the rent of the pastures in the "pago de Tlalhuapa," as was customary. The first mention of it in an inventory occurred in 1718, when nearly 40 acres of it were in cane. The mayordomo's account of 1768 notes that part of an aqueduct carrying water to the "campo de Tlalhuapa" had recently been constructed, and this supply must have supplemented that already in existence by at least 1718.

The rental price was based on an assessed value of the property of $450, contained in the terms of a censo agreement between the Indians of Jiutepec and Bárcena, administrator of the plantation (T1965). This value yielded an annual rent of $22-4-0, the amount paid from at least 1707 to 1851. Although Alamán remarked in the early 1850's (L417 E8) that the land supported merely low-grade pasture, he was referring to its unimproved condition, responsible for the low rental. In fact, at the time he wrote it was within the permanent fence surrounding the canefields, and had been used at one time to raise coffee.

Important events in the late seventeenth- and early eighteenth-century history of Tlalhuapan are described in L27 V48 E9 D2, containing a lawsuit begun in 1707 by the Indians of Jiutepec against Cristóbal Mateos, then administrator of Atlacomulco. Although the testimony was conflicting — so directly contradictory that it is impossible to identify the owner — the facts leading to the suit may be stated simply. Mateos dismantled some boundary markers and moved into 58 varas of the 55 acres, occupying only part of the tract. The Indians, claiming ownership since "time immemorial," had no documentary proof of ownership other than an amparo obtained from the Audiencia in 1688. The defense was assumed by Estate lawyers, who based it largely on the assertion that exhibition of title was necessary to prove ownership, implying that land lacking such title could be dealt with as Mateos had attempted, and stated furthermore that an amparo from the Audiencia was insufficient proof of ownership in the Marquesado, where title could be gained only from the Marqués. The lawyer of the Indians, in words described by the Estate lawyers as expressing "manifiesto desacato," replied sharply that the Indians had owned the land before the Marquesado existed, and that the "derecho de dominio" of the Marqués applied only to Spaniards, adding that Indians should not be required to show title, the statements of witnesses being sufficient in such cases.

No decision accompanies this document, but an introductory note states that the plantation was to pay the rent while the decision was pending: "manteniendo al Pueblo en la posesion que alego disfrutar y mientras justificaba la propiedad que aun no ha llegado el caso de probarlo." The decision must have been in favor of the Indians, since later in the century Atlacomulco paid them $24 per year for the 55 acres.

Tlalhuapan next became a source of friction between the hacienda and Jiutepec after Lucas Alamán attempted to buy it from Jiutepec in the mid-nineteenth century. Until this time, Atlacomulco had paid the annual rental of $24, the land was within the fence surrounding all the canefields of the plantation, and apparently no questions had arisen concerning it since the early eighteenth century. As soon as purchase was discussed, however, questions arose concerning the size of the piece, its boundaries, and whether or not it had water rights. Valero, prefect of Jiutepec, summarized the case (L417 E8) in the early 1850's and concluded that it was unlikely to be favorable to Jiutepec, inasmuch as the pueblo had no papers, its boundaries were unknown to even the oldest people, and Atlacomulco had made so many improvements — fences, diversion of water from Río Chapultepec — that one had to assume that the plantation had rights to the land. Estate officials, on the other hand, were less certain of ownership since documentary proof was lacking on both sides. Alamán (*ibid.*) was willing to buy the land but objected to the high price of $1,000 asked by Jiutepec, stating that the one-half caballería had no water rights, was merely pasture land of low quality, and that even $300 or $400 would be too much to ask for it. However, in view of the prospective advantage to be gained by escaping from the machinations of certain "malibolas" whose principal aim he implied to be the creation of disagreement between pueblos and haciendas over land ownership and boundaries, Alamán felt that the hacienda should pay $1,000 for the land, even though a search had revealed no documents either in Cuernavaca or Jiutepec concerning it. The area certainly was open to question, for although there seemed to be general agreement that it consisted of one-half caballería, in fact the area contained between the two boundary markers called Tranca de Macaca and Rincón del Patrocinio, finally agreed to be the limits (L417 E2), must have been much larger than that. Figure 5 shows that if the boundary is projected westward from El Patrocinio to the point where it intersects a line projected north from Tranca de Macaca, an area of about 300 acres is outlined. This figure is nearer the estimate made in 1760 by Alarcón, who drew an irregular figure of different shape, labeled it "Tlalhuapa" (see Fig. 4), and stated in the description accompanying the map that the land belonging to Jiutepec contained slightly more than 200 acres (L27 V48 E3).

It is difficult to see what the plantation had gained by purchasing under these conditions the land called Tlalhuapan. Even after the 1856 decision prohibiting review of the case, at least some of the inhabitants of Jiutepec objected to the sale when it was again reviewed about ten years later. Although Alamán felt that fixing boundaries was necessary to avoid contro-

versies with neighbors, in fact the sale did not overcome ill-feeling in Jiutepec. As in many other cases where ownership of property came into dispute in Morelos, the question seemed never to be settled, since recourse might be had through appeal to a number of courts and clear titles were few in number.

Rancho de Guimac. Atlacomulco made a major acquisition of land in 1791 with the purchase of the neighboring ranch called Guimac, a property whose history is so eminently typical of the histories of many other properties in Morelos that it is worth reviewing.

In 1643 Pedro de Haro Bravo, vecino of Cuernavaca, applied to the Crown (not to the Estate) for permission to use approximately 210 acres of land that had belonged to the depopulated village of San Mateo Sacualpanapa. He offered $20 per year (censo perpetuo), adding an initial $60 cash payment as the result of a competitive bid. His bid was accepted in 1644, and he sold the land in June 1649 to Blas Ocharte, the first in a series of such sales — many following brief tenures of only four to six years — that continued until purchase of the property in 1735 by the very wealthy Joseph de Movellán y Lamadrid, Regidor Decano and Contador de Menores y Albazeazgos of Mexico City, lessee of Atlacomulco, and lessee of the meat supply of Cuernavaca. In 1791, when he was twenty-five years old, the grandson and heir of Movellán wanted money to establish a capellanía, and hence took the offer of the Board of Directors of the Estate to buy the property. It had been rented by the Estate from the Movellán heirs since 1773 for pasture for the livestock of Atlacomulco, primarily to overcome the difficulties posed by shortage of grass in April and May at the end of the dry season. The livestock were removed from Guimac sometime in June, when the rains usually began, in order to begin plowing. Thus the property was used for pasture for about fifty years before its purchase in 1791 by the Estate. The Board offered $3,400, itemizing the value in the following way: for land, fences, and water, $4,242; subtract censos worth $600: final offer, $3,400.

Guimac continued to be used as pasture for the livestock of Atlacomulco, with only a small part, approximately 20 acres, rented after 1879 by the State of Morelos for use by a regional agricultural school to be established in the pueblo of Acapanzingo (L417 E24 contains a copy of the rental contract). The rental agreement ran for nine years, and the cost was $50 per year. At present, much of the area of the former ranch is irrigated and used to grow rice and cane.

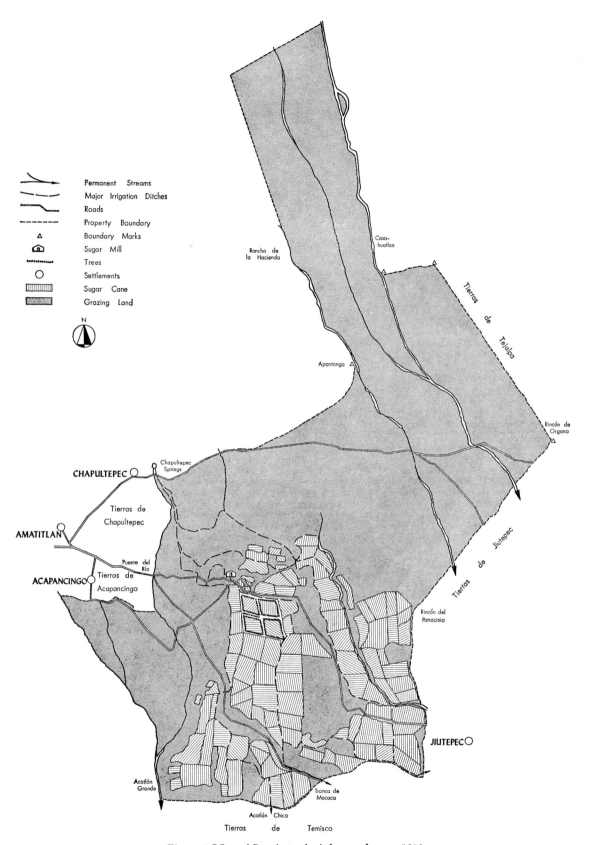

Legend:

Permanent Streams
Major Irrigation Ditches
Roads
Property Boundary
Boundary Marks
Sugar Mill
Trees
Settlements
Sugar Cane
Grazing Land

N

Rancho de
la Hacienda

Casa-
huatlan

Tierras de Tejalpa

Apantongo

Rincón de
Organo

CHAPULTEPEC

Chapultepec
Springs

Tierras de
Chapultepec

AMATITLAN

Puente del
Río

ACAPANCINGO

Tierras de
Acapancingo

Tierras de Jiutepec

Rincón del
Patrocinia

JIUTEPEC

Acatlán
Grande

Tranca de
Macaca

Acatlán Chico

Tierras   de   Temisco

Figure 5. Map of San Antonio Atlacomulco, ca. 1850

33

Estancia de Mazatepec. About 25 kilometers southwest of Cuernavaca Hernán Cortés established a ranch that came to be called Mazatepec. In the inventory of 1549 it is called Atelinca and described as "entre los pueblos que se dicen Mazatepeque e Micatan." The best description of the location of Mazatepec is found in L93 E17, a document that describes the leasing in 1690 from the Estate of what was then called the hacienda de ganado mayor Mazatepec to Captain Francisco López del Castillo for $500 per year. The document includes a description of the boundaries that shows the estancia to have been nearly coextensive with most of the land claimed in the late nineteenth century by the sugar plantation San Salvador Miacatlán (Observatorio Maps 2627, 2628).

The major physical features of the ranch were a shallow lake called Guatetelco, part of whose shore consisted of a rather extensive and swampy plain; part of the alluvial deposits of a river to the west; and, covering most of the area of the ranch, a gently rolling surface that supported only pasture and some temporal on its rather thin soils. The most important of these physical features must have been the swampy plain bounding the lake, particularly well developed on its southwestern corner. It was this feature that made the land important in supplying livestock to the Cortés mill and to the carnicería of Cuernavaca.

The property was never regarded as part of the entailed estate, and was listed by Luján (L260 E5) in the early 1600's as one of the bienes libres of the Marqués that could be sold. There is no information concerning its acquisition, and it ceased to be associated directly with the plantation sometime in the early seventeenth century.

The separate accounts of Mazatepec for the years 1589–91 (L262 E3) indicate the general nature of its operation. Livestock was received from the Cortés ranches in Tehuantepec and pastured at Mazatepec, whence they were sent as needed to the Taxco mines, the carnicería at Cuernavaca, and Tlaltenango and Atlacomulco. After Mazatepec ceased to be under the administration of the Estate in the early 1600's, the lessee or purchaser continued to supply the plantation with cattle and mutton. For example, the plantation account of 1620 shows that money was spent to buy livestock from Mazatepec. As time passed, the ranch became more closely identified with Temixco, Acatzingo, and finally Miacatlán, sugar plantations south and southwest of Atlacomulco, as well as more exclu-sively with the carnicería at Cuernavaca. In the complex concurso de acreedores that began in 1630 (T1725 E1) Mazatepec is included in the inventory of properties that had Temixco as nucleus, although no titles were presented and its status — whether rented or on a long term censo lease — is not described. It appears that this property, an important part of the economic network of the first and second Marqueses, lost its importance to the Marquesado as a result of the bankruptcy proceedings of the early seventeenth century; it does not appear in plantation accounts or as rental property associated with the plantation after 1624. Thus, although it is mentioned in a list of Estate properties of 1639 (L239 E21) as being "junto al dho ingenio," the description continues with the remark that it had been used for raising livestock but was currently "despoblado" and rented for $250 per year to Juan López Morgado of Temixco, who had contracted to supply meat to Cuernavaca. It was rented to Andrés de Egurén, and after his death in the early 1680's it was taken over by Francisco López del Castillo for $200 per year (L148 E1; L148 C2). Sometime before 1690 López del Castillo had established a trapiche at a ranch called Acacingo (Acatzingo; L93 E17), also owned by the Estate and adjoining Mazatepec. Together with Mazatepec, Acatzingo subsequently became part of San Salvador Miacatlán.

Nineteenth-Century and Multiple-Use Problems

In the nineteenth century a principal aim of the Estate administration, an aim that both stemmed from and intensified the existing problems with neighboring pueblos, was to establish a fence around the pasture land lying north of the irrigated canefields. As a result, for the first time there was produced a bounded, fenced, surveyed, and mapped property. Rows of trees were planted along part of the boundary (they still exist and can easily be traced on aerial photographs). However, merely placing boundary markers and drawing a map of the property were not sufficient to settle the question of ownership of the fenced land.

The major outline of the nineteenth-century history of ownership problems is found in the work of Lucas Alamán, much of whose effort in the period from about 1835 to his death in 1853 was directed toward these problems. In the series of published letters from Alamán to the Marqués, the earliest reference to difficulties with neighboring pueblos occurred in a letter of 27 June 1837 (*Documentos*, Vol. 4, p. 382), in which he stated that persons from neighboring pueblos were squatting on land customarily regarded as belonging to Atlacomulco, and that he had ordered one of the lawyers retained by the Estate to go to Cuernavaca to look into the matter. Both land and water, particularly the ill-defined boundaries of the plantation, were mentioned several times in 1838 as sources of difficulties that increased the problem of renting the plantation, owing to the time the renters had to spend on matters not directly related to sugarmaking (*ibid.*, pp. 398, 40). The matter had become so serious by July that Alamán himself went to Cuernavaca to investigate and resolve the problems, but his efforts were unsuccessful.

A gap in the published letters extends from 1839 to 1847, and after the latter date they contain no mention of land questions until 1849 (*ibid.*, p. 491) when Alamán wrote that there would be unexpected expenses owing to the need to buy rather than rent La Huerta and the Acatlanes. After this purchase, the only land rented for the plantation was some belonging to the pueblo of Tintepec, and in the following year of 1851 (*ibid.*, p. 199) its purchase was arranged. Finally, all the land used by the plantation belonged to it, and, according to Alamán, few plantations of tierra caliente shared this feature. His next move was to have the boundaries legally marked (*ibid.*), after which at least two maps were made of the plantation and some daguerreotypes taken of various parts of it. Figure 5 is redrawn from an uncatalogued map in the Archivo General that I assume to be the second map and the only one that Alamán approved.

Use of Pasture Land. By far the greatest amount of legal action of the nineteenth century concerned pasture land lying north of the irrigated canefields of the plantation. There were problems with all of the neighboring Indian pueblos concerning this type of land, with the exception of the pueblo Santa Catarina on the northeast, subject to the Villa of Tepoztlán. Although these problems had many aspects, principally they derived from conflicting but simultaneous use of nonirrigated land for temporal by Indians and others on the one hand and for grazing by plantation livestock on the other. Conflicts occurred because both uses started simultaneously at the beginning of the wet season, and it was too costly for either party to erect fences to keep the stock from the crops. The Indians held that the lands were ejidos, belonging to the community and open to use for grazing or planting by anyone, whereas the plantation wanted guaranteed access to even the

very scanty grazing that these thin, dry, rocky soils afforded, and no controversy over crop damage. The point that this type of land was the major source of controversy between the plantation and the neighboring pueblos in the nineteenth century may be worth emphasizing; by contrast, the irrigated land produced little controversy except in respect of the value of the pieces that the hacienda wished to purchase, as we have seen. Pressure for the use of the land for temporal came from both Indian members of the communities and non-Indians who had obtained, as they thought, rental leases from the Indian communities. From the point of view of the Indians, the population increase of the past two centuries had brought about the need for more temporal; from the point of view of the hacienda, the lack of other grazing lands and the increase in stock numbers required access to the pastures.

At a technical level, the controversy was encouraged by faulty records and a lack of records. Thus, a comparison of two maps of the plantation that must have been made within a very short time of each other in the mid-nineteenth century reveals that in many important details they differ remarkably from each other: in shape of areas shown, orientation of boundaries, and lengths of segments of roads or boundaries. Even the representation of such conspicuous physical features as some deep barrancas differs between the two maps, and there is striking confusion in the naming of some of the boundary markers, showing that it was not possible to obtain an accurate map of even major features to guide planning and discussion. Few of the Indian communities seemed to have any papers that even suggested ownership. Probably most of the communities had never had a delineation of their lands, or could submit no better evidence than early sixteenth-century listings of boundary markers with names that had fallen into disuse.

On the west, problems arose with two neighbors, Chapultepec and Acapatzingo, both separated by fence lines from the plantation rather than by physical features. Most of the land of the two other neighbors on the west, the plantation Temixco and the barrio San Pablo of Cuernavaca, was separated by a deep barranco from the plantation, and boundary questions never arose with either of them.

In the case of Acapatzingo, the amount in dispute was not large nor did it involve long discussion. Although there is no document stating the outcome of the controversy provoked by the survey of 1850 made for the plantation (L417 E3), it appears from the map which Alamán ordered made after the survey that he had backed down in the matter and conceded a small piece to the village. However, there was difficulty concerning Guimac, described in legal documents of 1854 (L417 E7). In this year the administrator Aguirre of Atlacomulco filed a complaint against the pueblo of Acapatzingo, demanding the return of the ranch to the plantation. The history of its temporary transfer by the plantation to the pueblo in exchange for the Acatlanes was discussed above. The five-year period was to end in 1854, but the administrator claimed that the terminal date was 12 January, the date of the auction, whereas the pueblo claimed it was 1 June, the date on which the transfer took place. The resolution of this conflict was achieved by an apparent compromise, in which the pueblo agreed to retire from Guimac if the plantation would allow it to pasture 150 head of cattle on the ranch. Since the administrator had already agreed to the latter use, both previously and in a separate compromise of another lawsuit with the pueblo concerning a fence between the two properties (L402 E36), the pueblo withdrew except to exercise its grazing right. Perhaps the feature of major interest in these cases was the contention of the administrator Aguirre (L417 E7) on 7 February 1854 that at least one of the lawsuits was arranged by only a few inhabitants of the pueblo who were using its name in an attempt to secure grazing rights for themselves alone. This charge is consistent with others made since the late eighteenth century by plantation representatives, both in lawsuits and correspondence; in many cases these interested persons were Spanish or mestizo rather than Indian.

Discussion with Chapultepec concerning land was much lengthier and more serious. In 1713, in the earliest manifestation of disagreement over ownership (L27 V48 E9 C3), witnesses reviewed the circumstances surrounding the ownership of all the land north of the old road to Tejalpa, shown on Figure 5 as claimed by Atlacomulco. It seems clear that at least since about 1670 the land had been regarded as ejido or common land belonging to Cuernavaca, part of which it had obtained about 1650 as the result of the bequest of a cacique who had lived in the pueblo of Aguatepec to the northwest. The land had been used for a variety of purposes. Part had been rented to individuals, apparently Spaniards or mestizos, who had raised temporal maize there; the oxen of Amanalco had been pastured there, as had stock brought from northwest of Mexico City for the meat

supply of Cuernavaca; a recent administrator of Atlacomulco, Diego Urtado, had rented the northernmost part of the pasture, apparently on his own account, and either he or the plantation had placed 1,000 head of stock on pastures north of the Tejalpa road. Nearly all of the witnesses in this case had been employed by the hacienda at some time, and the point was made by them that no renter of Atlacomulco had ever complained of sharing these pastures before Mateos raised the question in 1713, stimulated by the fact that the person who had contracted to supply meat to Cuernavaca was not only stocking heavily on the pastures but also changing the position of some boundary markers. Although no decision on his complaint is included with the copy of the document that I inspected, it seems clear from the statements of the witnesses that a large part of the land north of the Cuernavaca-Tejalpa road was in 1713 truly ejido land, owned by Cuernavaca. This conclusion is supported by the testimony of another witness, who stated that the pastures owned by the plantation and used by the oxen of Atlacomulco were south of the irrigated land (probably in the Acatlanes) and themselves were irrigated in the dry season. In the wet season, however, pasture land lying north of the plantation was used as well.

When a lawsuit was begun in 1756 concerning part of this land (*ibid.*), the administrator of the plantation asserted that the parts in question — only a small part apparently of the land north of the Cuernavaca-Tejalpa road — had never been owned by the Indians, implying that in fact it was Estate land, but that the Indians used it for temporal. Three characteristics of this case, in contrast to the earlier presentation, were that the amount of land claimed by the Indians was much reduced, much more specific information was given concerning its location, and it appeared to have much more the quality of a discussion concerning where the boundary should lie than a contest concerning a large block of land. This change is curious in view of the testimony given in 1713 that much of the land was ejido.

Tejalpa. Difficulties with the pueblo of Tejalpa involved the location of the boundary from El Patrocinio to Rincón del Organo, as well as shared use of the land on the Atlacomulco side of the boundary. Boundary problems with Tejalpa began at least as early as 1760 in the course of a boundary check described in L341 E14, containing the objections of the Indians of Tejalpa to the location of two segments of the line. They later objected, and more specifically, to the locations of the boundary markers set by the hacienda in the survey of 1850 (L417 E3). Partly in order to overcome these objections, Alamán moved to buy Tlalhuapan from Jiutepec in 1850, inasmuch as Tejalpa was subject to Jiutepec; it was felt by the Estate officials that the government of Jiutepec might be persuaded to convince Tejalpa that it should accept the boundaries as stated in the survey of 1850. Indeed, a promise by Jiutepec to attempt such persuasion was written into at least a draft of the Jiutepec-Atlacomulco agreement of 1850 (L417 E8). It was also stipulated that Atlacomulco allow the residents of Tejalpa to use part of the land on the Atlacomulco side for temporal and limited grazing without charge, as had been customary (L417 E8). I have found no record of the outcome of this disagreement, but in 1865 and 1866 the hacienda was "invadida, ocupado por la fuerza [del] pueblo [de Tejalpa]," a reference to what were described as five pieces of land previously granted to Atlacomulco that were along the boundary challenged in 1850 (L417 E10). The court (juez de primera instancia del cantón) decided in 1867 that since all the Tejalpa witnesses were residents of the pueblo, whereas only two of the Atlacomulco witnesses lived on the plantation, the case of Tejalpa was not proved under the provisions of a current law that specified qualifications of witnesses. As a result, he ordered that all the land be returned to the hacienda, citing yet another current law; in addition, Tejalpa was to pay costs for its "temeridad" and to lose the corn planted (L417 E10). Tejalpa raised the question in the Cuernavaca courts again in 1875 without success (L418 E19). These difficulties with Tejalpa continued in spite of the fact that, according to the lawyer of the pueblo, the titles of Jiutepec on which Tejalpa depended were inadequate, Atlacomulco had the better case, and the matter should probably be dropped (L417 E11); the judge agreed, and each party was ordered to pay his own costs. Even after this decision, in 1877 (L147 E21), residents of Tejalpa destroyed fences and planted corn within the disputed area; on 8 December 1877 the Juzgado de letras de Cuernavaca ordered the Tejalpa residents to leave and also to pay costs (*ibid.*).

The change in frequency and intensity of litigation over land following Independence appears to have had as a partial cause, from the point of view of hacienda owners and managers, incitement by mestizos and perhaps some Spaniards called by Alamán "malibolas." For example, in November 1814 Calleja, then manager of Atlacomulco and a person of superior analytical powers and a certain amount of objectivity, reported the failure of the Alcalde de segundo voto of the Ayuntamiento Constitucional of Cuernavaca to send Indians to work at Atlacomulco; Calleja remarked that the Alcalde had spent his life on sugar haciendas — indeed, he owned one — and hence knew that such plantations depended on the continuation of the modified system of repartimiento labor that had persisted through the eighteenth century. Apparently the Alcalde did not fear the loss of his power to enforce the rules of the system; rather he had gone so far as to tell the Indians, according to information given by one of their governors to Calleja, that "todos seran iguales y no tenian obligacion" to work on haciendas (L278 V48 E11). The Alcalde's motives were explained thus by Calleja: the Ayuntamiento Constitucional had assumed the management of community land, with the result that questions had arisen concerning land rented by Atlacomulco that had never before been a source of controversy. Calleja felt that the Ayuntamiento might take the land from Atlacomulco to rent it to one or several members of the Ayuntamiento, of which the Alcalde was a member. As Calleja remarked, simply changing the renters was of no advantage to the Indians. The obvious implication is that the non-Indian and part-Indian members of the Ayuntamiento saw opportunity in political change in New Spain, but were more interested in transferring ownership or use to themselves than in community rights.

When Alamán attempted to establish boundaries and ownership of parts of the plantation in the 1850's, numerous examples occurred of what might be called, from the point of view of plantation owners, obstructionism by persons holding public office in pueblos and towns. The feeling of the plantation class is well shown in the letter of an Estate representative present at a meeting held to discuss purchase of land from Jiutepec, when he discussed the behavior of the síndico propietario Sedano, the principal representative of the pueblo: "Así como hay autómatas cuyas máquinas les hacen movar la cabeza verticalmente para manifestar que sí, el Síndico llevaba instrucciones para movarlas únicamente en línea orizontal para decir a todo que no. . . . (L418 E3)."

Alamán stated clearly on at least two occasions that the struggle for land had strong racial overtones. In a letter of 13 March 1848 (*Documentos*, Vol. 4, p. 466) he wrote that the Indians of Sochitepec had attacked the ingenio Chiconcuac, to rob and also to seize the land, as well as to kill "toda la gente decente." Help was sent from all the neighboring haciendas to the besieged mill, and the American commander in Cuernavaca sent troops to the pueblo of Sochitepec, but, Alamán added: "cuando el ejército norteamericano se retire, mucho me temo que las revoluciones de este caracter se repitan, y que quedemos en mucha inseguridad. . . ." (*ibid.*). On 13 May 1848 he wrote that as soon as the Americans left civil war would begin and that "tomara el caracter de guerra de castas entre las varias que forman esta poblacion, y siendo de ellas la menos numerosa la blanca sera la que habra de perecer y con ella todas las propiedades que le pertenecen" (*ibid.*, pp. 470–71). In the face of unrest and the deep pessimism it engendered, the faith in the efficacy of placing boundary markers seems nearly pathetic, particularly since the locations of nearly all the markers were contested by the neighboring Indian pueblos in 1850 in the course of the first survey ever undertaken. Yet Alamán persisted, stating that as soon as some of the boundaries were in doubt, the plantation would lose the peace with which it had lived with its neighbors (L417 E8). Curiously enough, although the frequency and intensity of litigation increased in the nineteenth century, the peace of the past was less real than imagined.

Acquisition of Water

Since the water used for the mill at Tlaltenango could not be transported because of physical problems to the fields at Tlalcomulco and the new mill site there, the acquisition of water for the latter involved different sources. For the mill at Tlaltenango the major source was the springs of Santa María, the source of the river that flows along the west side of Cuernavaca, and for Atlacomulco it was Chapultepec Springs, one of the sources of the river that flows along the east side of the city, separating it from the mill at Atlacomulco.

Most of the information concerning the rights to the water used at Tlaltenango is contained in L282 E4, a collection of documents describing a suit brought by Isabel de Ojeda, widow of Serrano de Cardona. In 1529 her husband had obtained a lease in perpetuity from the Indians of Cuernavaca and the estancias of Tetela and Istayuca, north of the city, of land and water for Axomulco, at $240 annually (see Fig. 4). In effect, the water rights involved apparently only the springs of Santa María and the river to which they were tributary, since there was little other water available. Shortly after these rights were obtained Cortés built Tlalte-

nango on land adjoining Axomulco, and downstream from it. Although Tlaltenango began as a trapiche powered by mules, not requiring water power, some water was needed for irrigation near the mill, and the Marqués was allowed to use the excess from Axomulco for irrigation but not for power. The amount was not specified, but vagueness concerning both amounts and delivery is characteristic of such situations in Morelos; the affair became complicated by the fact that the second Marqués not only converted the trapiche to an ingenio, but also bought one-seventh of the Axomulco plantation from the founder's widow, Isabel de Ojeda. In 1553 agents of the Marqués also assumed the management of the mill, moved all its equipment to Tlaltenango, and diverted all the water not used for irrigation at Axomulco to power the mill at Tlaltenango and irrigate the fields there. According to testimony gathered in the case described in L282 E4, Martín Cortés bought an interest in Axomulco specifically to get the water, which was absolutely necessary for the improvement and expansion of Tlaltenango. He was apparently stimulated to do so by decrees of the Audiencia (auto and auto de revista, 1551, and sentencia definitiva, 1552) enforcing the earlier agreement between Hernán Cortés and the founder of Axomulco, particularly the provision that the water was first to go to Axomulco and then to Tlaltenango. With the ingenio at Tlaltenango thus crippled, purchase of part interest in Axomulco must have seemed the only way to keep the mill operating, and strong pressure must have been applied to Isabel de Ojeda to induce her to form a company with the Marqués in Axomulco. Isabel de Ojeda obtained a judgment in her favor in 1564 when she petitioned the Audiencia to force Martín Cortés to reestablish the mill at Axomulco, a judgment repeated in 1567 and again in 1570, by which time the lawyers of the Estate could plead that the difficulties posed by sequestration of the Estate had made reestablishment impossible. Axomulco was never rebuilt, and I have found only one later document referring to its water: in 1712 a reconnaisance was made of the use of water from the river, necessary because someone had applied for a grant of the water associated with the site of Tlaltenango to power a wheat mill in the vicinity of the ruins. The petition was opposed by the renter of Atlacomulco, Cristóbal Mateos, on the grounds that the water was used by the fieldworkers of Atlacomulco, most of whom still lived at the Tlaltenango site, to irrigate plots on which they grew food for their subsistence, and that the rental agreement between him and the Estate would be void if the grant were made because the water was clearly part of the plantation that he had rented; he added, but rather lamely in view of the physical difficulties and expense, that he was thinking of diverting this water to the site at Atlacomulco. Tlaltenango thus still possessed in the early eighteenth century water rights that must have been largely unused.

The development of water rights at Atlacomulco proceeded simultaneously with development at Tlaltenango, since the fields at Atlacomulco became at an early date a principal source of cane; before cane was planted there, mulberry trees had been grown under irrigation somewhere in Atlacomulco (L285). About 1540 water from Chapultepec Springs was diverted at its source by an aqueduct about a league long to the fields at Atlacomulco, and another much longer aqueduct, at lower altitudes, was under construction in 1549 (*ibid.*). No information is available concerning the acquisition of rights to this water, nor does it appear that the rights were disputed before 1768; other information available is primarily about delivery of the water to the mill and fields at Atlacomulco, dealt with in the next chapter. In 1768 (L161 E4) the Indians of Chapultepec took some water from a spring called Guadalupe, formerly used by Atlacomulco; I have not found the details of this case, which is merely described by title in L161 E4 as a legal expense of that year. After Independence, however, it became necessary to formalize the diversion of the water, which was done on 23 December 1842 by agreement with the inhabitants of Chapultepec (L417 E3).

Summary

As a result of the events described in this chapter, the cane-growing enterprise of the Marqueses del Valle became, by the early seventeenth century, a very different producing unit than it had been at its beginnings. Instead of consisting of scattered fields or groups of fields separated by Indian land, it had become a single compact unit with a single cluster of irrigated land and an adjacent pasture. Dependence on the distant estancia of Mazatepec ceased. Although some land continued to be rented from Indian communities until the nineteenth century, the purchase of a large tract called Tlacomulco in the early seventeenth century gave title to a sufficient nucleus of irrigable land. Most of the acquisition of land in the colonial period was peaceful, a characteristic that depended on the depopulation that affected Morelos in common with the rest of Mexico. After Independence disputes over land arose that centered on the nonirrigated pasture land north of the plantation that had been used for temporal by the adjacent Indian communities.

The process of acquisition of water rights is obscure. Apparently no titles to the water used by Atlacomulco exist; the use of water by Tlaltenango, abandoned in mid-seventeenth century, depended on the fact that the prior user, Axomulco, was incorporated within Tlaltenango in the sixteenth century.

V

Field Management and Yields

IN CASES where records exist, it is tempting to try to calculate the yields obtained on colonial plantations, for purposes of comparison. Naturally, there are difficulties, and the source of the difficulties here, as elsewhere, is inadequate data: for example, information concerning the production of molasses is generally missing, or the data in many cases refer to an isolated year whose representativeness may be open to question. In this kind of effort, a principal aim should be the establishment of probabilities for a wide range of yields, in order to evaluate the uncertainty of the operation and responses to the uncertainty. But adequate runs of years must be very rare, and one settles for much less —more or less accurate estimates of yields in given years, with the weather largely unknown. Another approach, the comparison of expectable yields as stated in handbooks (Barrett, 1965), offers some insight, but cannot portray the interannual variation that must have dominated plantation operations; even in the case of an irrigated plantation, where the vagaries of weather may be partially overcome, yields vary from one year to the next. In spite of the difficulties, some results may be achieved with the data at hand. The purpose of this chapter is to present some estimates of yields — and a clear sense of their limitations — as well as an examina-

tion of the complicated interdependence of some factors that affect them.

Irrigation Practice

Water Need in Morelos. In order to understand irrigation practice and other aspects of field management in Morelos it is necessary to discuss the water needs of plants there and how these may be met through natural sources and irrigation. This may most conveniently be done by analysis of the water budget, following the method of C. W. Thornthwaite.

The essence of the Thornthwaite method is to obtain by means of empirical relations between average monthly temperature and the loss of water from irrigated surfaces estimates of the amount of water that would be evaporated from the soil surface and transpired by plants if there were no limit to the supply; this quantity, called potential evapotranspiration, is then compared with the amount naturally available from two sources, precipitation and moisture stored in the soil, to obtain actual evapotranspiration. It is assumed that different species do not transpire at greatly different rates, and that 30 centimeters adequately represent a regional average of moisture stored in the soil.

Figure 6A presents graphically data published by C. W. Thornthwaite Associates (p. 380) for Cuernavaca. The curve of potential evapotranspiration remains more or less steady and at its lowest level — about 2.3 inches or one-fifth acre-foot of water per month — from mid-October to mid-February, owing to relatively low average monthly temperatures. Temperatures, hence also potential evapotranspiration, rise in March and April to a peak in May before the onset of the summer rains, when increased cloud cover reduces daily maximum temperatures, leading in turn to a fall in mean monthly temperature in June. Both temperature and potential evapotranspiration fall slowly

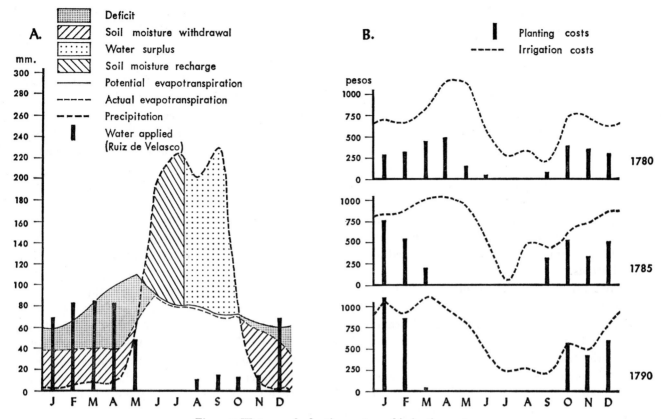

Figure 6. Water need, planting costs, and irrigation costs

from the peak in May until the end of the rainy season in October. The annual march of potential evapotranspiration is characterized by relatively small amplitude, since it is primarily a function of temperature and the latter changes little in Cuernavaca from season to season.

Water Supply in Morelos. Since scarcely any rain falls in the six months from November through April, soil moisture is the only natural means of satisfying water need, yet is inadequate to do so: from late October until April the deficit increases month by month. After the rains begin, and there is a large excess of moisture income over need, the transpiration needs of plants are satisfied, the soil is refilled to the 30-centimeter limit, and excess moisture appears as runoff, mostly in the months of August and September. Although the average annual potential evapotranspiration (935 millimeters) is less than average annual precipitation (1,025 millimeters), the curves clearly show that the marked concentration of precipitation in a single period only four months long results in a water deficit during the other eight.

In colonial Morelos the deficit was made up entirely

by diverting water from perennial streams and springs rather than by constructing storage dams that would permit significant extension of the irrigated area by impounding both runoff in the wet season and the perennial flow. Probably the failure to build storage dams resulted not only from the nature of the stream courses and their flow, but also from the fact that two contrasting types of irrigable surface exist in the region. The streams in many places are deeply entrenched in narrow barrancas with nearly perpendicular sides, and peak flow is often torrential. Where flood plains are wide, they offer preferred sites for irrigation rather than water storage. In addition, the presence of a line of perennial springs at the base of the escarpment invited the construction of diversion rather than storage dams.

The two contrasting types of surface of Morelos are associated with soils of differing moisture-holding capacities, owing not only to textural differences but also to differences in depth. The heavier lowland soils are much deeper than the upland soils — some of the lowlands were swampy before drainage in post-Conquest times — and the water table in many depressions is much nearer the surface than it is in the upland. These

41

facts emphasize the unreality of assuming the same storage capacity for both types of surface. The upland soils may have a storage capacity of 10 centimeters or less; 30 centimeters may be a realistic figure for the lowlands.

Water Need and the Timing of Agricultural Operations. In Morelos, crops and other vegetation give easily perceptible indications of water deficit and adequacy, and irrigation need is obvious in a plant such as cane, whose cycle of growth covers twelve months or more. Nevertheless, it is useful to compare the annual march of the components of the budget with the annual cycle of agricultural activity in order to discover changes in field routine that may be related to changing evaluations of the possibilities and limitations offered by seasonal rhythms.

Figure 6B shows the annual course of irrigation labor costs and planting labor costs for three years in the late eighteenth century, by which time agricultural operations seem to have become highly routinized and in phase with seasonal climatic changes. The sharp decline in irrigation costs between May and June expresses a rapid decline in the need for irrigation because of increasing precipitation. In general, July, August, and September were slack field seasons, both for irrigation and for planting; in one year, 1785, when fairly large-scale planting began earlier than usual in September, irrigation costs had begun to rise in the previous month, perhaps in anticipation of the planting needs and suggesting that the dry season was earlier than usual. It appears from these curves that by the eighteenth century transpiration needs were judged to be met quickly and nearly fully with the onset of the summer rains, although in August, September, and October, when planting had begun and rainfall still provided a large surplus of moisture, some irrigation was considered necessary to ensure sprouting (Ruiz de Velasco, 1937, pp. 348–56; Fig. 6A). Approximately the same small amount was supplied in November; in December, the amount applied was slightly in excess of potential evapotranspiration, suggesting that soil storage had been exhausted and that 30 centimeters are an overestimate of the amount of storage, or that cane requires more water than many other plants. In the remaining months of the dry season the amount applied remained fairly near potential need, dropping rapidly in May with the onset of the rains. If it is kept in mind that Figure 6A sets forth average conditions within which are concealed interannual variations in the amount and

timing of precipitation, it appears that by the late eighteenth century (Fig. 6B) the general outlines of the water application routine bore a close relationship to water need.

Development of the Water Conveyance System. Although the acquisition of water rights was dealt with above, it remains necessary to discuss delivery of the water to the mill and the fields. By 1549 (L285) separate aqueducts had been constructed to the sugar mill and the wheat mill in operation at Tlaltenango, probably using the Axomulco diversion at the springs of Istayuca. In addition, by the mid-sixteenth century two aqueducts were built to carry water to the fields at Atlacomulco from Chapultepec Springs (L257 E14), and one or both were enlarged about a century later when the mill was moved to Atlacomulco and the need for more water for power became apparent (L269 E42). At this time a third aqueduct was added to irrigate fields in the southwestern part of the plantation. In the mid-eighteenth century, when the renter Tomás de Avila Romero decided to convert from animal power back to water power, a second period of rebuilding and extensive repairs was necessary, as a result of which the value of aqueducts was approximately one-third of the total value of the plantation (T1965). As in the mid-seventeenth century, three aqueducts were in use, and this pattern of conveyance changed little until the mill was finally abandoned. The major nineteenth century improvement was the construction of a larger diversion dam below Chapultepec Springs to obtain the greater flow necessary to power a turbine that had replaced the old water wheel (L27 V48 E4).

Quantity of Water at Atlacomulco. I have found only one reference to the amount of water that belonged to Atlacomulco. Alarcón, writing in 1760 (L27 V48 E3) and describing what must have been only the source at Chapultepec Springs, wrote that "las referridas tierras . . . gozan de el beneficio de las mejores aguas, que ay en esta jurisdiccion, como asi mismo de hallarse debajo de cercas, de que se hallan resguardadas de todo el ano," putting the amount of water at 58 surcos, which according to his calculations were distributed over 984 acres.

The Irrigation Routine. Ruiz de Velasco (1937, p. 217) claimed that Cortés and other hacienda managers installed in Morelos the Arabic system of irrigation without important modifications, and that it differed from irrigation practice in the rest of Mexico (*ibid.*, p. 219) except in Puebla and Oaxaca (*ibid.*, p. 226). He published a diagram (*ibid.*, p. 223; here reproduced as

Fig. 7) labeled "Sistema de riegos Hernán Cortés en el Estado de Morelos," implying little change since its introduction until at least the late nineteenth century. The complexity of the system is nowhere even hinted at in plantation accounts or in other descriptions of Morelos.

As Figure 7 (A–D) suggests, irrigation consisted of a sequence of four major stages distinguished by successive reductions in the number of tributary irrigation ditches (regaderas; *c* in the figure). The water moved from the main ditch (apantle; *a* in the figure) into the regaderas or apantles, thence into the feeder ditches (tenapantles or contra-apantles; *b* in diagram E) at intervals of twelve furrows (surcos). Each group of twelve furrows was called a tendida (E), and one man was assigned the task of regulating the flow of water in three tendidas.

In the first irrigation (7A) the water entered the furrows from the tenapantle and left by the first regadera (Ruiz de Velasco, 1937, pp. 222–23); this was a relatively light application of water, sufficient to wet the fields to encourage sprouting. In one or two weeks, depending on the weather, the second irrigation (7B) began, with every other regadera eliminated by joining the segments of the furrows that the regaderas interrupted, an operation called mancornar. The riego de dos apantles continued at intervals of from 8 to 14 days for a total of two to four months, after which the number of regaderas was again halved to form the riego por mitad (7C). This operation again doubled the length of furrow through which the water traveled before it moved from the field, and at the same time the amount of water was doubled while the time elapsing between applications was halved. The riego por mitad continued for one or two months, or until the internodes of the young plants began to form; after this time the plants would be so large as to make troublesome the elimination of the regaderas, so the one remaining in the middle of the field was eliminated to begin the riego de punta, when the water at last ran directly through each furrow from tenapantle to achololera. The amount of water applied in each irrigation then reached its maximum, approximately 176 cubic meters per hectare. The use of the tendidas now became more important: for a given daylight period all the water coming to each group of three tendidas tended by one man might be diverted wholly into one of the three, and in the following into another, so that the frequency of application reached its maximum of once every three days. As local

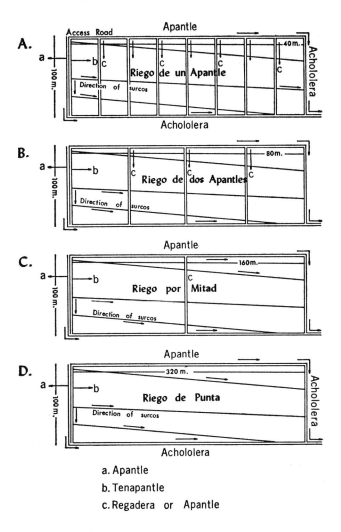

a. Apantle
b. Tenapantle
c. Regadera or Apantle

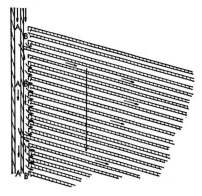

Tendida:

AAA Apantle

BBB Tenapantle or Contra-apantle

Figure 7. Plan of irrigation ditches

usage had it, in this stage "el agua duerme," that is, circulated in a tendida for six or twelve hour periods, soaking the fields.

Ruiz de Velasco pointed out the advantages of this system by contrasting it with the Hawaiian, which had shorter (10–20 meters) furrows lacking drainage ditches. The longer furrows (the shortest in Ruiz de Velasco's example was 40 meters long), and the successive changes in their length allowed better adjustment of supply to need; finally, provision of drainage ditches assured better drainage.

Advantages of Irrigation. Irrigation is useful in Morelos at times of frost, since watering is the only way to counteract its effects; the colloquial Mexican expression for frost damage, that the plant se seca, suggests this remedy immediately. Furthermore, irrigation enabled most planters to feel that application of fertilizer was unnecessary, since the silt carried by the water provided at least some nutrients for the plants (*ibid.*, p. 358; Lewis, p. 144). Recognition of the fertilizing effects of irrigation led to use of the latter even in the rainy season, because at that time the plants were large enough to benefit markedly from the practice (Ruiz de Velasco, p. 358). The other major advantage of irrigating was that rats were killed and ants kept in check (*ibid.*, p. 341). It is possible that the monetary benefits of irrigation, expressed in rodent and ant control and in removing the need to fertilize, outweighed its costs, since the amounts of money spent on the salaries of the ditch-tenders were only a very small percentage of the total labor bill.

Cultivation

Only two instruments were available for ground preparation and for weeding and cultivation, the plow and the indigenous coa. As an alternative to the coa, the hoe seems to have been used only in the sixteenth century.

The Plow. The Roman or criollo plow was in use, and usually five parts were distinguished in inventories: cabeza, reja, telera, timón, and mancera. Of these parts the handle (mancera) is first mentioned only in the late eighteenth century, possibly because it need not have been a separate piece of wood from the sole (cabeza). Both beam (timón) and sole were made of oak (encino), and it is possible that the handle was made of the same material, or, if it was one with the sole, might be made of either iron or wood. The share (reja) was made of iron, and the telera, the pin that fastened the sole to the plow, might be made either of iron or wood. Both of the latter were made locally, although in the late eighteenth century most of the inventories contain references to rejas castellanas, and nearly all the shares are called Castillian, perhaps a reference to design rather than to place of manufacture. The beam and sole were made in the carpentry shop at the plantation in the sixteenth century, but later some were bought from Indians.

After Independence, the 19½ moldboard plow, the disk plow, and the arado carro or wheelplow came into use in Morelos (Ruiz de Velasco, 1937, p. 191). The new types were not adopted until 1850 at Atlacomulco; the plows were English, required mules for their operation, and were much better than the old arados del país that were still used on the plantation (Alamán, *Documentos*, Vol. 4, p. 593). The disk plow was used at Atlacomulco at least by 1905 (L420 E15). Luis (Noriega) wrote that the 19½ moldboard plow was generally used in Morelos for the first and second plowing, and was drawn by two oxen, plowing to a depth of about 20 centimeters (p. 5). The old chisel plow was still used in the 1880's (*ibid.*, p. 22) after the cane was about two months old in order to redistribute earth, and Atlacomulco had twenty shares for such plows in 1884.

As was the case with so many of the different kinds of tools used at the plantation, very small numbers of both plows and parts of plows were kept on hand until the late eighteenth century. The case of plowshares is illuminating: until the inventory of 1767 between 4 (1767 and 1625) and 21 (1567) shares were on hand, but in the inventory of 1777 ninety-seven rejas castellanas were listed, and subsequent inventories show similarly high numbers available. In addition, through the inventory of 1693 fewer than two dozen plows or soles were on hand, but in the next inventory of 1718 there were 130 soles alone, even though only 20 plows were operative. The disproportion, probably indicative of the fact that the plowsoles wore out or broke within a short time, persisted until the inventory of 1777, when 44 operative plows were available, but maximum numbers of plows were kept in 1791 and 1799 (72 and 63), after which only smaller numbers appeared.

Hoes. The hoe (azadón) was used at Atlacomulco, and its mention in inventories extends from 1581, when there were 74 present, until 1847, when there were only 2. Most inventories between these two list fewer than half a dozen. The azadón may have been used in the mid-sixteenth century for ground preparation, at a time when many workers were available, but by at least 1625 they were no longer used for such a purpose, since 6 of the 10 then listed were used for the aqueducts, presumably for cleaning them. By the late eighteenth century they seem to have been used only to prepare clay for the purging process.

Coa. Instead of relying on the hoe for weeding, the workers of Morelos favored the indigenous coa, used constantly in Morelos for "mullimiento, escardas y riegos," as Ruiz de Velasco (1937, pp. 178–79) put it. These instruments had a short wooden handle and an iron blade of variable form that the "herreros de las haciendas . . . confeccionaban a gusto de los trabajadores" (*ibid.*, p. 192). The coa was used mostly in what were called beneficios of irrigation and particularly for weeding; so strongly was the tool identified with the latter work in place of a hoe that Ruiz de Velasco (*ibid.*, pp. 178–79) claimed that in Morelos the word "escarda" was scarcely used, the substitute term being "de la mano de coa." However, throughout the late eighteenth century both terms seemed to have enjoyed almost equal frequency of use in accounts and other documents from Atlacomulco. The usefulness of the tool extended beyond simply weeding to include a certain amount of earth-moving to encourage root development. Luis (pp. 21–22) was critical of the instrument, pointing out that it was used only in Morelos in cane cultivation and was extremely inconvenient because it required stooping, but he added that no other instrument had been devised that could be inserted between the cane plants to put soil around the bases to improve root development. He noted also, however, that for the third weeding the arado del país achieved the same purpose of redistributing soil.

The evidence in the inventories suggests that the Indian workers were expected to bring their own coas to work, at least from the sixteenth to the mid-eighteenth centuries, since there were hardly ever more coas at the plantation than there were Negro slaves available to use them; indeed, the inventories of 1611, 1613, and 1618 state that the approximately 30 coas in stock were for the use of Negroes. As with other tools, the inventory of 1777 marked a departure from the earlier practice: with hardly any slaves at the plantation, 92 coas were listed, in 1791 there were 187, and in 1884 as many as 105.

Plowing. The functions of plowing were numerous enough to require a rather complex system of successive treatments of the land to destroy the old cane-

stocks, loosen the soil, and prepare regaderas and furrows. Four initial plowings were followed by preparation of the seedbed. Ruiz de Velasco (1937, pp. 192–96) and Luis (pp. 5–7) show that the first four plowings were made in different directions, and that the subsequent preparation of the seedbed was so difficult and important an operation that it was supervised by the mayordomo himself. It was in this last step that the irrigation courses were laid out by eye, first the furrows, then the regaderas. The latter had to be both parallel and at carefully measured intervals, for they had the additional function of providing two of the four sides of the area of piecework called the tarea.

Maintenance of Soil Fertility. Ruiz de Velasco (*ibid.*, p. 358) mentions the beneficial effect of irrigation water, whose silt and dissolved nutrients were generally considered to be adequate fertilizer in Morelos. He added (*ibid.*, p. 284) that the sludge and silt (azolves) deposited along the sides of the irrigation canals and long neglected finally began to be carted to the fields and were discovered to increase yields sufficiently to pay for the costs of collection and distribution. Haven (pp. 244–45) wrote that the clay used in purging was also carried to the fields for whatever benefit it might confer, and Ruiz de Velasco (*ibid.*) felt that inclusion of the cartroads within the cultivation cycle might be beneficial, since they were well fertilized. The use of a system of fallowing was described by Henry Ward (Vol. 2, p. 65), who stated that only one-fourth of the cane land belonging to most Morelos estates was cultivated annually, a practice that yielded what he and a companion experienced in the British West Indies industry regarded as abundant crops. It was mentioned above that dunging was not practiced at Atlacomulco, although corrals made ample dung available; Luis (p. 4) wrote in the 1880's that redistribution of the manure from the corrals was unprofitable.

Growing

Cane Varieties. Fortunately for this analysis, experimentation with cane varieties did not begin in Morelos until about 1840, so that modifications in practice and changes in yields cannot be interpreted as functions of varietal change. Even when Ruiz de Velasco was writing in the 1930's (p. 25), the criolla variety brought by Cortés was still the most common variety in the state, yielding an abundant juice rich in sugar. The variety had great vitality, had not degenerated, and was so disease-resistant that cane diseases were always insignificant in both Morelos and Puebla; however, it had the disadvantage that it was extremely sensitive to extremes of heat and cold. It was a thin-reeded plant, about 3.5 meters high at maturity. On 3 February 1849 Lucas Alamán wrote that he had planted some Tahitian cane, called in Morelos habanera, and that he believed that it was more productive than the criolla; he proposed to use the former exclusively if the hacienda soils proved suitable, but there is no further information on the outcome of the trial (*Documentos*, Vol. 4, p. 491).

Planting. Probably several interrelated aspects of plantation management and sugarmaking account for the changes that occurred in the dates of planting shown in Figure 8. It can be seen that the planting dates became stabilized in the late eighteenth century, a time when many other practices of the plantation came under review. Another feature of importance here is the coincidence of the shift with a markedly decreased reliance on Negro slave labor, which meant that plantation rhythms had to be adjusted more closely to the seasonal needs of the labor pool by placing the slack period in the temporal, when Indian fieldworkers and other free laborers were planting and cultivating their own food crops. In fact, as is shown in Chapter V, the change may have been advantageous to the plantation, inasmuch as cane cut in the wet season produced more molasses and less and weaker sugar than cane cut in the dry season, and harvesting dates were, of course, dependent on the time at which cane was planted; at the same time, the change caused a shift from less costly dependence on seasonal rainfall to more expensive irrigation at the beginning of plant growth. Making planting and harvesting simultaneous and limited to the dry season was suggested in the Instrucción of 1770 (L298 E50?), where the writer remarked that, when planning fieldwork, the administrator should make certain that whenever a field was harvested there should be a plowed field ready in which to plant some of the cane harvested. This directive records another change in practice that became standard sometime in the late eighteenth century, the exclusive use of plant cane for milling rather than depending on a combination of plant and ratoon.

In planting, sections of cane perhaps eighteen inches long were placed in and parallel to the furrows, their ends overlapping until the outermost nodes of adjacent sections were very close to each other (Ruiz de Velasco, *ibid.*, p. 213). The planted sections were covered with

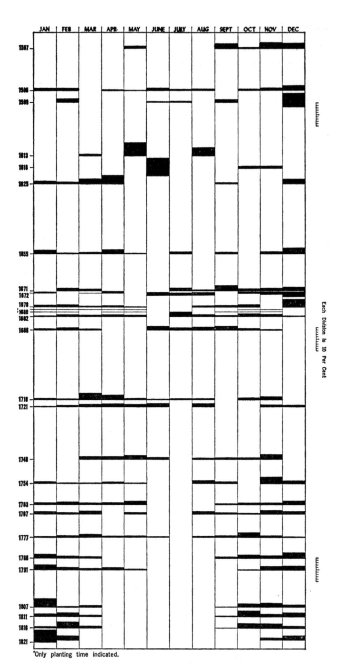

Figure 8. Planting regime at the hacienda, 1567–1821

earth taken from the tops of the ridges between the furrows, leaving the former truncated, and as soon as possible the first irrigation was given so that the cane would sprout.

Weeding and Beneficios. Although these operations were of crucial importance to the development of the cane, Ruiz de Velasco gives them scant attention on the grounds that they were too well known and obvious in their results to merit attention (*ibid.*, p. 235). For this reason, it is necessary to rely on the fuller description given by Luis, which suggests that the two activities were closely associated; basically the beneficios consisted in the gradual reduction of the truncated ridges separating the furrows where the cane had been planted, moving the soil to form finally a ridge where the furrows had been. At a maximum, Luis suggested three hand weedings and two light plowings. The Atlacomulco accounts show that two each of weedings and beneficios were performed.

Harvesting

Age at Which Cane Was Harvested. The age at which cane is ready to harvest depends on a number of factors, including the variety, the weather, the season in which it is planted, whether it is plant or ratoon cane, and the timing and thoroughness of plowing and weeding. Latitude and altitude are also sources of variation in maturation from place to place. In areas like Morelos, where irrigation is necessary but planting and harvesting may follow seasonal rhythms, complications in scheduling may be introduced because of variability in weather as well as dependence on ratoon crops, which generally mature more quickly than do plant canes.

Because of the complicated dependence of the length of the maturation period on a number of factors, some variable and some not at the same place, it is impossible to give more than a general indication of the age at which cane was ready for milling at the Cortés plantation. Current experience in Morelos cannot be a guide, either, since the colonial variety of cane is no longer grown there on an appropriate scale. However, enough of the inventories indicate the state of cane at different ages to enable us to establish at least a range of months at which cane was ready.

It is important to note that second and third ratoons (resoca and quarta soca) were important only in the early seventeenth century, and by the last quarter of the eighteenth century the dependence on first ratoons had almost ceased. If we discount the evidence of the relatively great age of the oldest plant cane (18 to 22 months) in 1811 and 1812, when difficulties in harvesting were experienced from the Revolution, it is apparent that the growing cycle was reduced from the late sixteenth century to the late eighteenth century. From 1567 to 1618, when all except one (1606) of the inventories were taken in the dry season when milling was also going on, the oldest plant cane fell within the age range of 16 to 28 months. At the upper extreme of the range (26–28 months), the cane was being cut at the time the inventory was taken, but in 1807 and 1812 cane even younger (22 and 25.5 months) was regarded as too old to be worth milling. The youngest plant cane described as ready to cut in 1567 was 13 months old, whereas in the eighteenth century cane even a month older might not have received its final cultivation. In 1721, plant cane 17 months old was described as likely to be lost if not cut immediately, whereas a few years before (November 1718; L30 V57 E3) it was necessary to mill it at 13 or 14 months. These extremes make it appear that most plant cane was regarded as mature 15 or 16 months after sprouting by the early eighteenth century, and that plantation managers were less willing than they had been previously to mill cane that had been standing for more than 17 months.

Recognition of the suitable maturation age seems to have preceded working out a rational planting schedule by six or seven decades, since a relatively simple solution to the complex problems of scheduling seems to have been reached only in the late eighteenth century; this solution is best illustrated by data from the years 1807, 1811, 1816, and 1821 (Fig. 8). Some planting began in September, but nearly all cane was planted between the beginning of October and the end of March. If a month is allowed for the time between planting and sprouting, and maturation was achieved within fifteen months of sprouting, then cane planted at the beginning of September would be ready at the beginning of the second January following, and if the mill was prepared to accept some slightly less mature cane, milling could be finished in June.

The record of inventory ages, discussed above, combined with the record of planting dates (Fig. 8) and the knowledge that stabilization of harvesting dates occurred in the late sixteenth century, indicates that optimization of the timing of planting and harvesting proceeded slowly, with the three major aspects of the problem — age at maturation, dates of planting, and time of harvesting — treated separately from one another.

Harvesting Methods. The detailed annual accounts of the period from 1581 to 1625 show that cane was cut in relatively uniform amounts daily, so that usually about twenty cartloads were delivered to the mill; it is shown in Appendix D (p. 113) that this unit was called a tarea and may have been approximately equal to an acre. The work of cutting was performed by gangs of laborers, including Negro slaves, and the tool they used for cutting may have been like the common machete in use today, although the word does not appear until the inventory of 1625. Previously the words applied to the tool were despingador (desburgador?) and cuchillo, but there can be no doubt about the identity of the object described, since in many cases the statement is added that it was used to cut (cortar) or clean (limpiar) cane.

Total Area in Cane

Information contained in inventories permits calculation of the area in cane at various times; Appendix D discusses the problems posed by changes in the units of measurement. The results are summarized in both Table 11 (p. 129) and Figure 3 (p. 19). The area under cultivation saw its first peak in 1549, when 274 acres were in cane or plowed in readiness for planting, although the total area available for cultivation was much greater than that figure; until the last two decades of the century smaller amounts were cultivated. The death of Fernando Cortés and restrictions on the labor supply resulted in a fall in new planting and a catastrophic drop in the production of sugar. The total area under cultivation, however, was maintained at about 200 acres through reliance on first, second, and third ratoons. A low proportion of plant cane was apparently maintained until the inventory of 1655, when the cultivated acreage exceeded 300 acres, and in subsequent inventories the area of ratoon cane decreased. The cultivated area remained between 250 and 350 acres from the 1650's until the 1750's, after which the available figures show that these extremes increased by 100 acres or more, with two peaks (1763 and 1807) exceeding 500 acres, and consistently higher percentages of plant cane. Not until 1851 were the peaks of 1763 and 1807 exceeded, when the record for area under cultivation was established under the stimulus of record prices. Sugar production quotas were established by at least the early 1870's by the state of Morelos, whose legislature was controlled by plantation interests; Atlacomulco, with a quota of about 325 tons of sugar and 725 tons of molasses (Haven, p. 245, says 375 tons of sugar; Morelos, Secretaria General . . ., *Informe General* . . ., 1872), probably maintained a cultivated area of about 600 acres, as in the inventory of 1884.

Mean and Median Field Size. Average field size in acres can be obtained from data in twenty-nine inventories falling between the first (1549) and the map of about 1850 (Fig. 5), a period of 300 years. In seventeen, well over half of the inventories, the average field size fell between 10.0 and 11.9 acres. After the inventory of 1688, or for more than half the period, average field size did not fall below 9.0 acres. Median and average field size corresponded closely, beginning with the inventory of 1671. Many of the fields in different inventories fell within the range 9.0–12.9 acres, indicating that the average and the median for the late seventeenth and the eighteenth century were representative of the size of numerous fields; in 1721, for example, 14, or nearly half of the 30 fields, were between 9.0 and 12.9 acres in size, while in 1791 slightly fewer than half, or 16, of the fields were in the narrower range of 9.0 to 11.9 acres in size. In the case of most ranges in the late seventeenth century and through the eighteenth century, the extremes were made up of as many as five small fields less than an acre in size, called machuelos (or more commonly machos in the sixteenth century), or at the other extreme of a few large ones exceeding 25 acres; the latter consisted in many cases of large tracts that had just been plowed and might be divided when planted into units nearer mean field size. For most of the inventories from 1671 to the map of 1850, nearly all of the fields fall within the range 5–15 acres.

Yields per Acre

Yields of sugar per acre of cane harvested at the Cortés mill can be calculated for many years in the well-documented period 1581–1625, and for various years in the late eighteenth and early nineteenth centuries. A major source of difficulty throughout is the lack of records concerning the amount of molasses produced at the mill, which, as explained below, cannot be assumed in any year to have been approximately equal in weight to the sugar produced, as the old rule of thumb in clayed-sugarmaking would have it. The latter relationship may be satisfactory for use in describing long-term averages, but does not apply to day-to-day or even year-to-year yields. The result is that it is not possible to compare precisely yields per acre from this plantation with yields from modern cane plantations. In addition, a lack of knowledge concerning the age at which most cane was harvested, described above, allows only estimates concerning yields per acre and per unit of time, a formulation of yield necessary in comparing production in areas characterized by different ages of cane at maturity.

The first specific statement concerning expected yields dates from 1556 (L267 E26), when 1.5 suertes (8.25 acres; see Appendix D) were expected to yield

12,500 pounds of sugar (1,000 loaves), thus producing at the rate of 1,515 pounds of sugar per acre. The rate of yield may refer only to the top grade of sugar (blanca or white), or to all three types that were distinguished at this time, but since the white generally constituted about 80 per cent of the weight of sugar produced, the only alternative estimate is about 1,900 pounds per acre. In 1568 (L107 E3), when an attempt was made to calculate a reasonable compensation for frost-damaged cane on the basis of norms given by experts representing both sides in a lawsuit, it was stated that one might reasonably expect from a tarea of ratoon cane 1,000 pounds of white, 325 pounds of espumas, and 250 pounds of panelas, or 1,575 pounds of all kinds of sugar per tarea, about 1,433 pounds per acre. Since the amount expected from plant cane would have been larger, the norms of these two years appear to be in fairly close agreement.

All the actual yields that I have calculated for various years in the sixteenth and early seventeenth centuries (Table 12, p. 130) are considerably below these expectations. The range of the data for 37 years extends from 1,247 down to 404 pounds per acre, but the low figures should be regarded as the result of temporary disorganization of the mill and abnormal. Most of the remaining values are near 1,000 pounds of all kinds of sugar per acre, and we may assume that on the average the total yield of sugar and molasses per acre was about 2,000 pounds for the period 1555 to 1625, that with an average harvesting age of 18 months 1,333 pounds of both sugar and molasses were produced per acre and year, and that the data show little trend during this time.

However, some of the difficulty posed by merely doubling sugar yields to obtain total yields of sugar and molasses per acre are illustrated by results for four years between 1619 and 1623, when the weights of two of three classes of molasses are also available. Of the three types of molasses — the kind obtained because sugar crystals would not form or because molasses rather than sugar was desired, the kind obtained from the first draining of the pots, and that obtained over a longer period by claying — data concerning only the first two types are available. Interestingly, this information indicates that inclusion of at least two types of molasses tends to make total yields per acre more uniform even when the amounts of sugar produced per acre varied from 660 pounds to 1,014 pounds: three of the years had yields near 1,350 pounds per acre, including some of the molasses, and the year in which the greatest amount of sugar was produced (1621) had both the smallest amount of molasses and the smallest amount of total production.

Figure 3, showing trends in total production of sugar and total acreage in cane, indicates that the latter grew more slowly than the former; this suggests that yields of sugar per acre must have increased over time, an inference supported by data in late eighteenth-century accounts of the plantation, summarized in Table 14 (p. 130). These figures may be accepted with confidence, inasmuch as the accounts from which they were derived were set up in such a way as to indicate the loaves of sugar made from fields of known area and in many cases from cane of known age. Since at least the average weight of a loaf of sugar for a milling season or for the year is also available from the records of shipments made to Mexico City, all the information necessary to obtain accurate statements of yields per acre — in some cases, per acre and year or month — are at hand.

Although the average seasonal and annual yields per acre in the late eighteenth century occupy a range from 702 pounds in 1768 to 5,009 pounds in several months of the following year, most of the figures are well above the sixteenth- and early seventeenth-century yields of about 1,000; eleven of the nineteen figures in Table 14 exceed 1,500 pounds. In individual years, the yields obtained from individual fields varied widely. A good example of the latter variation occurred in 1779, when from seven named fields ranging in size from 5.5 to 19.9 acres, yields per acre were obtained that ranged from 753 to almost twice as much, 1,396 pounds.

In the mid-nineteenth century, the highest value expected was considerably less than the very heavy yield of 1769. In reviewing the results of the first grinding of 1848, Alamán wrote (*Documentos*, Vol. 4, pp. 473, 479) that 14.5 loaves, each with an average weight of 22.5 pounds, per tarea de siembra was a very good yield; in 1849 (*ibid.*, pp. 504–5) he predicted approximately the same yield, about 2,800 pounds per acre, from cane that had been planted in November. By the time this cane was milled in May 1851 it produced 2,735 pounds of sugar per acre. In 1851, the yield obtained in the first milling season was slightly less than the yield of the first grinding season of 1848, about 2,800 pounds per acre. Both seem to have been without precedent in the sixteenth and seventeenth centuries, but not in the eighteenth.

VI

Sugarmaking Techniques and Equipment

IN THE West Indies and the Spanish and Portuguese possessions of the New World, two principal types of sugar, clayed and non-clayed, were made before the introduction of modern sugarmaking techniques. The non-clayed types of sugar were most likely to be associated with the simplest facilities and the smallest scale of production, and were called in the English possessions muscovado and concrete; in Central America and Mexico, where the latter still constitutes an important place in the diet, concrete is called piloncillo, chancaca, panela, and panocha. In Central America it is still made in very small mills, most of which produce no more than a few tons annually; only in Nicaragua are there larger mills producing this staple in amounts of fifty or a hundred tons annually, or on a larger scale than that of the most ambitious panocha factories of the sixteenth and seventeenth centuries in New Spain. In Mexico large amounts of piloncillo are produced by modern mills in the form of cones weighing a pound or more; in Nicaragua the loaf is characteristically shaped somewhat like a brick, and the weight is about the same. In this type of sugar the molasses is incorporated with the grains to produce a hard and solid mass with a strong molasses flavor, a medium to dark brown color, and rather poorly developed granular structure — that is, there is no attempt even partly to refine the sugar

by separating the molasses from the grains. Muscovado, the other kind of non-clayed sugar, was much more important in its time than panocha as an export commodity, but is nowhere made today. It differed from concrete in being a soft mass of molasses and sugar. Muscovado was never important in Mexico, although the Cortés mill at Tuxtla shipped a large amount of its production in this form to Spain.

In the largest mills of New Spain and the West Indies the final step in the production process consisted of removing the molasses from the mixed mass of molasses and sugar that resulted from boiling to produce a conical loaf of sugar more or less free of molasses and known as clayed, purged, or white sugar. Since the separation of molasses from the grains could not be fully achieved by the techniques available, even the clayed sugar might have a light yellow color and a light molasses flavor from the thin coating of molasses over the grains. From the beginning, the sugarmakers at the Cortés mill and the other large mills in Morelos produced clayed sugar, and most of them were still making it, with only slight modifications in sixteenth-century boiling and purging techniques, when the Revolution began in this century in Mexico. Unlike panocha manufacture, the manufacture of clayed sugar has not survived technical and political revolutions.

The typical pre-Revolutionary sugar mills of Morelos relied on a process having three readily distinguishable stages: milling, boiling, and claying. The following pages describe the plan of the mill at Atlacomulco where these processes were carried on, after which the processes of manufacture are dealt with in their proper place in the sequence of sugarmaking.

Plan of the Mill
Only two plans exist, neither to scale, that show the functions of various parts and rooms of the mill. One of

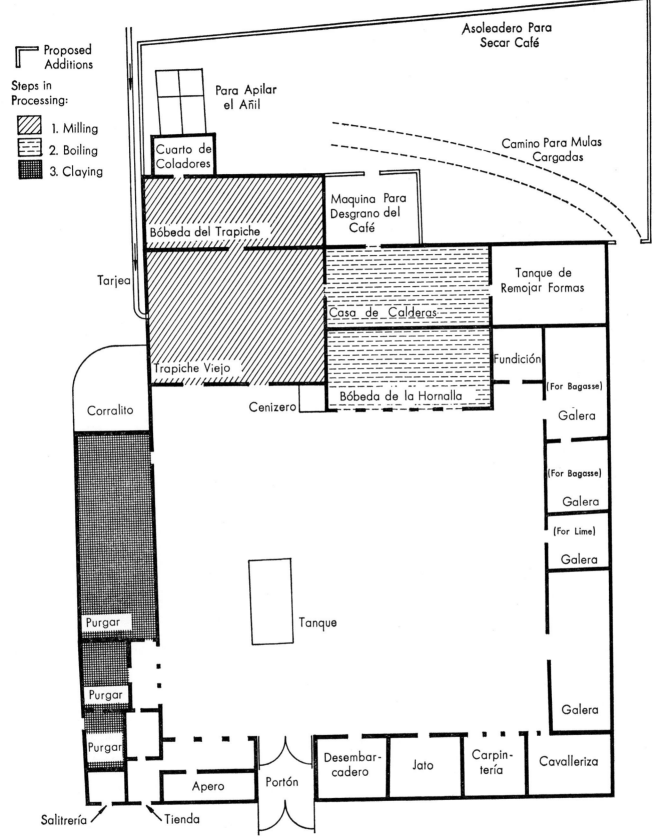

Proposed
Additions

Steps in
Processing:

1. Milling
2. Boiling
3. Claying

Asoleadero Para
Secar Café

Para Apilar
el Añil

Cuarto de
Coladores

Camino Para Mulas
Cargadas

Bóbeda del Trapiche

Maquina Para
Desgrano del
Café

Tarjea

Tanque de
Remojar Formas

Casa de Calderas

Trapiche Viejo

Fundición

Corralito

Bóbeda de la Hornalla

Cenizero

(For Bagasse)

Galera

(For Bagasse)

Galera

(For Lime)

Galera

Purgar

Purgar

Tanque

Purgar

Galera

Desembar-
cadero

Jato

Carpin-
tería

Cavalleriza

Apero

Portón

Salitrería

Tienda

Figure 9. Plan of the mill in 1824 (not to scale)

51

these was drawn by the mayordomo in 1824 (Fig. 9) in order to describe the location of planned indigo and coffee works; the other was published by Ruiz de Velasco (1937, p. 137; Fig. 10) and represents the mill about 1910. In addition, aerial photographs taken in 1959 show the plan of the ruins.

The first inventory available after construction of the mill at the new site in the 1640's dates from 1655, and begins, as do other seventeenth-century inventories, by describing the casa de vivienda, at that time reached by a wooden staircase. The living quarters were small, consisting merely of an anteroom with the staircase, and a larger room of the same width, apparently always poorly furnished (Calderón de la Barca, p. 376).

From the living quarters a door led to the upper story of the purgery, with the same dimensions as the lower — 40 varas long and 13 wide. A new purgery had been added, adjacent to the old and identical in width, but 49 varas long. The combined measurements of both the new and old purgeries are very little different from the space shown as occupied by the purgery in Figure 9.

Either the mill house or the boiling-house was mentioned next in most inventories, and both occupied the same places as in Figures 9 and 10. The former was 33 by 16 varas, the same size as the space occupied by the trapiche in the plan; to its east was the boiling-house, and south of the latter were the furnaces. From 1655 to at least 1721 the size of the boiling-house remained 24.5 by 12 varas, so that its larger size at the present time indicates that about 6.5 varas of length were added, probably sometime in the eighteenth century.

In most of the seventeenth-century inventories references made to casas de molino imply that one of these casas was used to store bagasse, so it is possible that the

space used for distilling in Figure 10 was occupied by the bagacera. By 1680 (L93 E7) a corral for storing wood had been added adjacent to the furnaces; this may have occupied the space of the northernmost of the galeras on the east side of the plant in 1824.

Several buildings were listed in 1680, in the following order, and it seems likely that they occupied the same spaces and positions as those indicated on the east and south sides of the patio: two rooms (casas) of adobe, covered with tajamanil; two jacales of stone and mortar (piedra y lodo), one used as an infirmary and the other to stable horses (cavalleriza), both also covered with tajamanil. In the plan of 1824 the stable is shown in the southeast corner, and there is a space north of it appropriate for housing the functions described in 1680. The inventory of 1680 also described a carpenter's shop and smithy, but without reference to their locations, as well as a kitchen (of piedra y lodo). Both jato and apero, shown on the south side in 1824, are mentioned in eighteenth-century inventories. A pottery (formería) appeared in 1721, as did two corrals west of the mill house and a small corral for chickens north of the purgeries. The dimensions of the chapel were given as 19 by 9 varas in the seventeenth and eighteenth centuries, but the one now in existence is larger, approximately 30 by 13 varas.

Thus by the 1720's it is possible to account for nearly all the space enclosed by the well-preserved exterior walls at the present time, with the exception of the patios north of the boiling-house and trapiche, which were added in the 1820's to process coffee and indigo, and the rooms to the west of the area labeled purgeries. It is possible that the latter space was occupied in 1678 by the thirty-four huts of Negro and Indian families, described in the inventory immediately after the contents of the church had been surveyed. Even the dimensions of the more or less ruined units within the walls are compatible with the dimensions of the large numbers of beams purchased and listed in inventories — primarily 8, 9, 10, and 12 varas long, or of the same size as the width of the rooms along the east and south sides of the main patio. It seems reasonable to conclude that the basic plan is that of the original construction of the 1640's, enlarged by additions on the north and west.

Although one might usefully plan alternative and more efficient arrangements of the rooms housing the various processes and activities occurring in the mill, the only obviously better arrangement would have

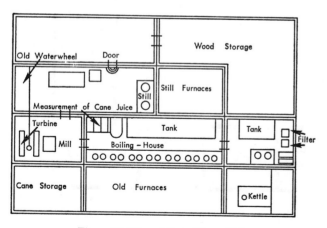

Figure 10. Plan of the mill, ca. 1910

more entrances and reduced crossflows. Such an arrangement would not realize what appears to be the major intent of the actual plan, that of maximum limitation of entry in the interests of security. The single entry suggests congestion of traffic, with wood and cane, the major bulky items entering the mill, moving through the same door. In fact, however, the patio is spacious, wood can be stored, and the number of cane carts entering daily was never so great as to cause inconvenience.

Milling the Cane

Although there are numerous French and English drawings of cane mills driven by water or by animals, I know of no published Mexican manuals that contain such illustrations. Sandoval published a plate called "Maquinaria de un ingenio de Azúcar en el Siglo XVI" which, although it does not appear to me to depict cane-milling equipment, is nevertheless useful because it shows an overshot wheel geared to three or four processing devices, indicating both ingenious applications of water power and very little difference between sixteenth-century Mexican water-powered machinery and early eighteenth-century water-driven machinery such as that pictured by Labat (Vol. 3, pp. 234, 246). Sandoval also published two diagrams entitled "Inventos para mejorar un trapiche de caña en 1818" (mule-driven) that are also very similar in their major features to the machinery in the plates published by Labat. Other than the plates in the work of Sandoval, the only information available concerning the mills at Atlacomulco is found in inventories and in letters from the mayordomos. The inventories, unfortunately, contain many words not found in modern dictionaries or now given meanings different from those they had in colonial times, with the result that it is impossible to reconstruct the major features of Mexican mills without relying heavily on the similarities between the machinery shown in Sandoval's plates and English and French drawings, using the much fuller information available concerning the latter to describe the general features of Mexican equipment.

The upper part of Figure 11 shows an animal-driven trapiche identical, except for the drive, with a water-driven cane mill illustrated below it. The illustration of the animal-driven mill is a nearly faithful copy of a plate in Labat's *Nouveau Voyage . . .* (1722, Vol. 3, p. 223). Labat's work refers to conditions in the sugar industry of the French islands in the last decade of the

seventeenth century, but his illustration of an animal-driven mill shows an apparatus very similar to the less detailed illustration in Ligon's work describing the sugar industry of Barbados nearly fifty years earlier. I assume that seventeenth- and eighteenth-century mills throughout the Caribbean and New Spain differed very little from these illustrations. Sixteenth-century inventories of Tlaltenango list three-roller mills that may have been of this type. In this vertical three-roller mill, a Sicilian invention of the mid-fifteenth century (Deerr, p. 536), the motive power was applied to the center roller of the three, all of which had their centers on the same axis. Cogs on the center roller engaged cogs on the outer two, with the result that they rotated in opposite senses and made it possible to achieve continuous double crushing of cane simply by feeding the cane between one pair of rollers and then passing it, once crushed, for a second crushing between the other pair. The novelty of this machine was the achievement of double crushing, hence more thorough extraction of juice, without the need for two sets of rollers or two sources of power. The advantage of the vertical arrangement is that juice runs more freely down the rollers than in a horizontal arrangement of the same three rollers, for in the latter case some juice pressed from the cane passing between the upper rollers would be reabsorbed by cane passing between the bottom two rollers. Easier flow of juice down the rollers was made possible by the adoption of shallow vertical fluting in the late eighteenth century, which also saw the adoption of the doubleuse (Dutrône, p. 104), a strip that guided cane, after its first crushing, between the second set of rollers and thus eliminated the need for one of the two attendants at the mill. It does not appear that the doubleuse was widely adopted, and Sandoval's Mexican trapiche of 1818 does not have one.

The wooden cylinders of the mills used at the Cortés mill were apparently from the beginning covered with sleeves of copper, perhaps also of bronze, but apparently never of cast iron. The wooden inner cylinders were made of wood called tepehuaje. Probably the metal sleeves were fixed to the wooden cylinders with nails, and the iron gudgeons on which their lower ends pivoted must have been prevented from turning independently of the wooden cylinders by means of flanges also attached to their lower ends. The gudgeons are partly visible in the upper part of Figure 11 at D, through slots in the bank of the mill where grease was applied to the bearing surfaces. The upper sets of gudgeons must also have been greased. The primary function of the slots at D, however, was to allow the introduction of wedges that regulated the spacing of the rollers to accommodate cane of varying hardness or thickness.

The parts wore out rather rapidly. Most inventories list several replacement rollers, some finished, some in process of being shaped, as well as sleeve and cog replacements, implying that the latter were fixed individually to the rollers. Most inventories list a dozen or more teeth available as replacements in the carpenter's shop, and in most accounts a bearing or two had been bought for the same purpose. The pivots at the bottoms of the rollers may also have required frequent replacement. Large quantities of tallow (sebo) were bought for the plantation, much of which was used on the bearing surfaces of the mill. Even so, these wore out rapidly: for example, in 1558 four metal bearings weighing almost 200 pounds were bought for the shafts of the "ingenio y molienda," to replace some that had been installed only two years before. In addition, the wood was subject to decay (L262 E2 D5; 1587).

The chassis of the mill had to be constructed of huge timbers to withstand the strains and stresses to which it was subject. The drawing of 1818 reproduced by Sandoval, which includes a scale in varas, shows that the four vertical posts had one dimension in excess of a foot, and were approximately six feet high. In Figure 11 the massiveness of the chassis is readily apparent. The uprights, braces, and crosspieces of the more powerful water-driven mills were larger than those of the horse-driven mills.

Source of Milling Power. The idea is widely held that use of the word "ingenio" implies that the motive power of the mill was provided by falling water, as contrasted with "trapiche," where mules or oxen moved sweeps that were connected with the central of the three rollers. In part, this interpretation of the words is supported by the distinctions made in the Mercedes collection of the Archivo General, where are found licenses for the construction and use of two different types of enterprise distinguished not only by the different names, but also by the different charges made for them. In addition, it was specified in many grants that if the mill was converted from a trapiche to an ingenio, a higher annual payment had to be made. There is no question that the ingenio was more valuable, and in general produced a greater annual value and weight of sugar.

In most sixteenth- and seventeenth-century inven-

tories of the Cortés plantation, the milling equipment is called a trapiche, whereas the entire processing establishment is referred to as an ingenio. It seems to be true that plantations producing the more expensive clayed sugar were called ingenios and likely also to use water for power, whereas those producing the cheaper panocha used animal power and were called trapiches. Almost from its founding Tlaltenango used water power, produced clayed sugar, and was called an ingenio, and the same characteristics were found in combination after the mill was moved to Atlacomulco.

Although it is impossible to reconstruct from the inventories the details of the Atlacomulco waterwheel, it seems certain that it was not horizontal (molino de bomba or molino de regolfo — Reti, pp. 388, 392). The use of the vertical rather than the horizontal is important in light of the evidence discussed by Reti that horizontal wheels were old, once widespread, "amazingly efficient," and well described by the Spaniard Juanelo Turriano about 1560, shortly after Tlaltenango was converted from animal to water drive.

Some scanty information concerning the waterwheel dates from the nineteenth century. In 1826 (L219 E3) a Scotsman named Glennie, apparently in the employ of a paper mill at San Angel, arrived at the mill to check the wheel and make some adjustment to increase its power. By this time both a large and a small flywheel had been added to the driving mechanism, and these were the parts that required adjustment. Later in the century (Ruiz de Velasco, 1937, p. 137) a turbine was substituted for the waterwheel, but in the inventory of 1884 a turbine was not mentioned.

There is a brief review of extraction and milling rates achieved by the three-roller vertical mill in Barrett's summary of performance standards of Caribbean mills (pp. 156–57). Given the information presented by Ligon, dealing with conditions of 1657 and a mill driven by five oxen or horses, it is possible that a milling rate of about 1 ton of cane per hour required only about 1/12 of an acre of cane to be harvested in each hour of daylight. This rate of supply is in agreement with evidence from the Cortés mill showing that about a tarea (approximately an acre) was cut daily; although cane was cut only in the daylight hours, the processing ran around the clock. The rest of the data show that it was not the mill alone that regulated the rate of production of sugar, but that the speed with which the juice could be processed in the boiling-house also was important. With the capacity of the typical seventeenth- or eighteenth-century battery, it seems likely that the water-powered mills available were capable of keeping up with boiling capacity, so that increased efficiency of the wheels and the mills would have required corresponding increases in the numbers or sizes of kettles in the boiling-house; to increase efficiency as well as the acreage harvested, the probable adjustment would be to install two batteries or separate processing facilities.

Pressing the Bagasse

After the cane had been crushed between the rollers, enough juice remained in the bagasse to justify further crushing in a press, of which the Cortés mill had two. The pressing may have been required for two reasons, first because some juice was left in the bagasse that might profitably be extracted, and, second, because it was difficult to burn the bagasse if it was wet. The use of the presses continued until the mid-eighteenth century.

Bagasse presses were used only in the Spanish colonies, and Ligon (p. 90) rather grandly explains the fact that the English did not use them:

Having passed [the canes] twice through, that is forth and back, it is conceived all the juyce is preft out; yet the Spaniards have a press, after both the former grindings, to press out the remainder of the liquor, but they having but small works in Spaine, make the most of it, whilst we having far greater quantities, are loath to be at that trouble.

Atlacomulco, at the time Ligon was writing, was about as large as the largest mills on Barbados.

Inventories taken before the 1750's, when the use of the press was discontinued, give too little information to permit its reconstruction. The most complete description is found in a document concerned with establishing the cost of a new press, described as the largest in the Indies, installed at Atlacomulco in 1654 (L238 E2). It was made in Mexico City and assembled on its arrival at the mill; carrying it to the mill and adjusting it required more than six months and the labor of more than fifty people. According to Governor Valles, under whose orders it had been built, the huge timbers were very valuable — the chassis had one member valued at $2,000, and the beam (palo) was worth $3,000. Its value was estimated in 1657 by two assessors, who agreed that it was worth the large sum of $11,000. The wood and labor cost $7,350 and 38 loaves of sugar; repairs were required when the press broke down after about three months, costing slightly more than $3,000.

The Boiling-House

Plan of the Boiling-House. Ruiz de Velasco's plan of part of the sugarworks (Fig. 10 above) shows the typical layout of a boiling-house, whether Spanish, Portuguese, French, or English. The room is much longer than wide. In most cases the kettles were arranged along one wall directly over the furnaces, which at Atlacomulco were below ground level, leaving the floor of the boiling-house itself near ground level. Ranged along the other side were tanks in which the volume of juice was measured, then transferred to another tank whence, as needed, it was transferred to the kettles.

Equipment. Although in Spanish the boiling-house was called the casa de calderas, possibly the English term "boiling-house" offers a more accurate view of the purpose of the activities carried on in it, since it contained also peroles, bacías, and tachas, other copper vessels in which juice was heated, in addition to the calderas, the largest of these vessels.

All of the calderas were constructed in the same way, presented in Figure 15 (p. 63) based on descriptions in inventories of 1682 and 1688. Each was of near conical shape, thus very different from French and English equipment, surprisingly deep (5 or 6 feet) in relation to surface area, especially since the purpose of the boiling was to reduce by evaporation the water content of the juice. Each consisted of a lower part called the fondo, directly exposed to the flames, and made of either molded or beaten copper, above which were added three tiers (andanas) of beaten copper plates (paños) held together with copper rivets (clavos). The andanas had different numbers of plates; in the caldera de melar, for example, the lowest or first tier had 6, and the second and third each had 12. The kettles were built up in this way, rather than formed in one piece, because only the fondo was directly exposed to the fire; to resist burning it had to be heavier and thicker than the upper parts. Savings of copper were made possible through this construction, since the fondo could be discarded after it became thin and the tiers of plates added to another new fondo or used for some other purpose.

In the 1870's Haven (pp. 243–44) wrote the best description that I have seen of the kettles used at Atlacomulco:

[The juice] passes into boilers about ten feet deep, four feet of copper, two of brick, and four of wood. The copper only holds the liquid; the upper part, opening widely, is for the froth and scum, good and evil, to disport in. The boilers under fire are filled to the brim with this bubbling, which is constantly skimmed by workmen, with flat skimmers half a yard across. . . . The syrup is taken to other boilers, where it is condensed yet more. . . .

This description is clear, simple, and reasonable, but it is impossible to say how well it describes the construction of the kettles in colonial times. The lower six feet of copper and brick may be assumed to be old, but the wooden construction of the top four feet I have never seen mentioned elsewhere, and in this respect it differs markedly from Ruiz de Velasco's slightly later description. Ruiz de Velasco (1937, pp. 247–48) disapproved of the conical kettles widely used in Morelos, and advocated shallow vessels with proportionately much greater evaporating surfaces.

Arrangement of the Battery. The purpose of the activities carried on in the boiling-house was to reduce the amount of water and impurities in the cane juice to the point where sugar crystals could form, a purpose achieved by heating the juice in large kettles such as those described. Wherever muscovado and clayed sugars were made, the juice was transferred through a sequence of kettles. Two major systems of arranging the kettles to form a battery existed by the early eighteenth century. The older, and the only one that was used in Mexico, consisted of a linear arrangement of kettles, each hung over a separate fire or, in a nineteenth-century modification, in pairs of kettles (mancuernas; Fig. 10) over separate fires. An alternative arrangement was developed in the West Indies, apparently in an effort to conserve fuel, and called the Jamaica train. In this arrangement, all of the kettles (usually five in number) were placed in a line over a flue leading from a single fire; here the kettles were graduated in size according to order of use or diminishing quantity of liquid, with the largest farthest from the fire, and the smallest or last to be used directly over it. The Jamaica train expressed the principle that as the process advanced the juice must become hotter. In the Morelos system, neither movement between kettles nor changing temperature is implied, although both were in fact part of the process.

The Atlacomulco inventories are unclear concerning the sequential nature of the operation conducted there. The difficulties may be seen immediately by comparing the information in Table 15 (p. 130) with Figure 12, showing the arrangement of kettles in a French West Indian boiling-house. The reduction in size as the operation proceeds is obvious in the French example, but in the Atlacomulco lists it is not clear which of the kettles was first in use.

There is information available that enables us to arrive at a more or less firm conclusion concerning the order of the named kettles, based on the meaning of caldo and melado. In the inventory of 1721, the kettle listed first is called "caldera . . . de resivir el Caldo," where caldo means untreated cane juice. Melado, derived from melar, refers to cooked syrup, and this fact makes the caldera de melar or the caldera de contramelar last in the sequence. The meanings of melado and caldo were well established by at least the early seventeenth century, when the mayordomo de Mendoza wrote as follows in reply to the Estate Controller's asking why he obtained only half the amount of sugar from his calderas de caldo that previous mayordomos had obtained (L268 E3 DB):

Las calderas de caldo que el da en su cuenta aver salido de la cana son del caldo que caya crudo en la primera caldera como salia del molino y que este cuando llegava a la ultima caldera donde estava en punto el melado para azer azucar avia consumido y mermado las dos partes y que si sus antecesores dieron mas açucar hecha de menos calderas de caldo seria porque no contavan las calderas de caldo crudo sino puesto en punto dando nombre de caldo a lo que era melado para hacer açucar y que esto se verifica por la cantidad de carros de cana que se molia en cada tarea y que lo que verdaderamente se llama caldo es lo que cae en la primera caldera que biene crudo del molino y prensas sin haver gozado de fuego alguno y no el melado que esta en punto para hacer el açucar. . . .

This statement, the only one I have found describing early seventeenth century practice, helps to establish the order of use of the kettles, as well as the fact that the syrup was moved from one to another in the course of boiling, and gives besides the proportion of expected loss through evaporation and skimming as two-thirds of the original amount of juice.

Clarification and Boiling. According to Antonil (p. 137), clarification was the first step to occur in the Brazilian battery, in the caldeira de meio. At Atlacomulco, as the name caldera de en medio suggests, this caldera — the smallest in the 1680's — was in the middle of the battery of five calderas. Impurities present in the untreated juice rose through the action of heat and added lye to form a scum, called in Brazil cachaça, in Mexico cachaza; Antonil wrote that this was skimmed off the surface and delivered outside the boiling-house by pipe to a trough where animals fed on it. In general, boiling

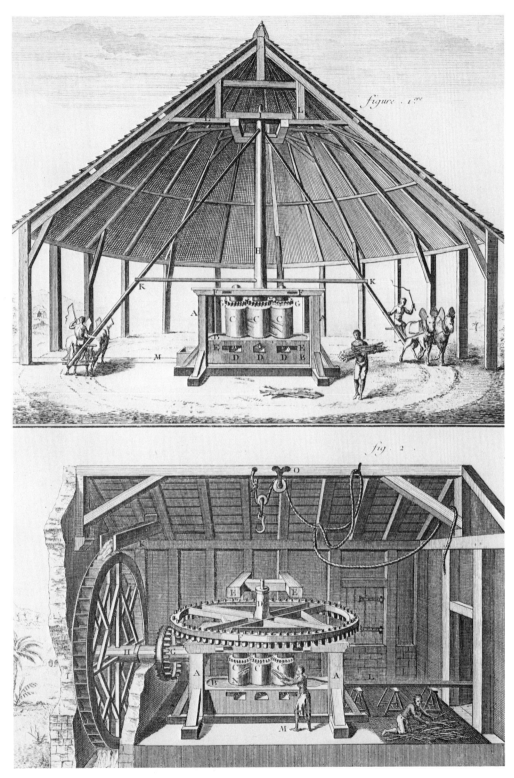

Figure 11. Animal-driven and water-driven mills

Figure 12. Eighteenth-century boiling-house

Figure 13. Eighteenth-century purgery

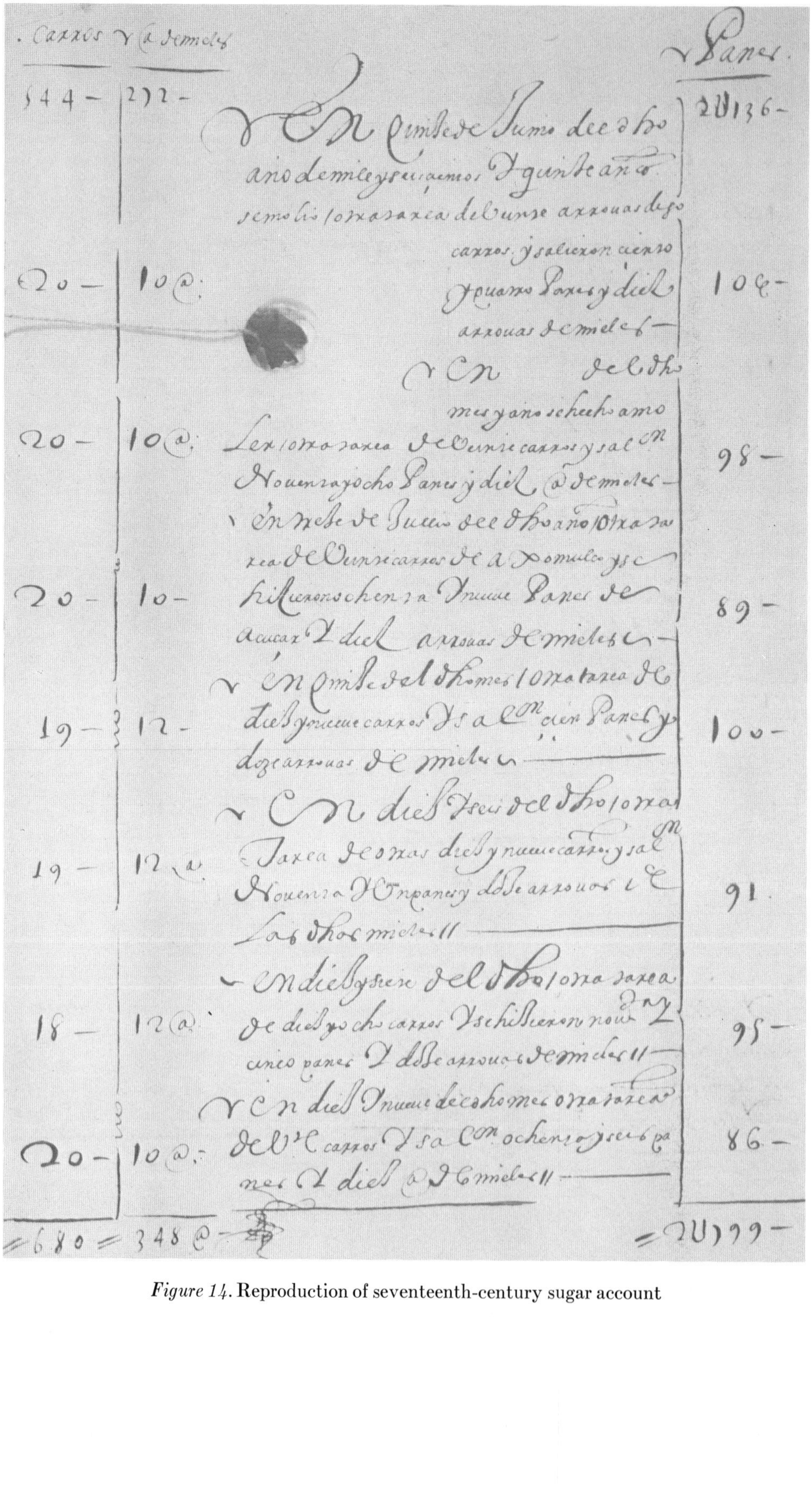

544 —   272 —   2U136 —

20 —   10 @:   10 @ —

20 —   10 @:   98 —

20 —   10 —   89 —

19 —   12 —   100 —

19 —   12 @   91 .

18 —   12 @   95 —

20 —   10 @:   86 —

= 680 = 348 @   = 2U199 —

Figure 14. Reproduction of seventeenth-century sugar account

was to be avoided in this first step because it had the effect of breaking up the scum and causing the risen impurities to be redistributed within the juice; unheated water might be added in order to prevent boiling (ibid.).

After about half an hour in the caldeira do meio visual evidence was expected to show that it was time to transfer the liquid: smaller and clearer bubbles rose to the surface, at which time it was transferred to the caldeira de melar to be heated for about an hour, with the same object of reducing the amount of impurities it contained (ibid., p. 138). The first skimmings of this copper, together with the later skimmings of the first, were transferred in Brazil to a parol where it was saved for the slaves, who used it to make garapa ("guarapa" in Spanish), a simple fermentation product with low alcoholic content; the later skimmings of the second copper were called in Brazil claros and there mixed with cold water to make an unfermented drink (ibid., p. 139). The caldo at this stage was sufficiently transformed in Brazil to be called by another name, meladura, and was transferred to another vessel called the parol do melado (possibly the equivalent of the Atlacomulco caldera de contramelar), where it was referred to as mel. From the latter it passed to a vessel called the parol de coar, passing through a cloth filter during transfer. Until at least the eighteenth century the use of such filters was standard at Atlacomulco, where they were made of jerga de Michoacán, and Labat recommended the use of filters (ibid., Vol. 3, p. 279); the English, however, were not so strongly inclined to take their benefits for granted.

It can be seen from the above summary of Antonil's work, which I assume to describe methods somewhat similar to those in use in Morelos, that only the first two coppers were adequately dealt with; his description breaks down in dealing with the last three paroles or caldeiras, leaving the reader without a clear picture of the rationale of their use. This is not at all the case with his French contemporary Labat, who expected something different to happen in each, and watched all six coppers for signs in the liquid indicating the proper timing of events. Antonil's may have been the confusion of inexperience; Labat enjoyed more than a decade of experience as administrator of plantations in the French West Indies.

Much clearer is Antonil's description (p. 140) of the differences in function between the larger caldeiras and peroles, on the one hand, and the smaller tachas on the other, which correspond, from the point of view of size, to the differences between the calderas and tachas at Atlacomulco: the calderas were used for clarification, and filtering of the juice took place as it passed from one to the next; in the tachas the final and much hotter boiling occurred that immediately preceded cooling and granulation. This distinction may also help to explain the great difference in size between the two: the Atlacomulco teaches in the 1680's and 1721 were hardly a tenth as large as some of the kettles, which suggests that the contents of the final kettle might be split between two teaches. The fact that the teaches appear heavier in relation to their capacities than do the kettles doubtless indicates that their sides were made thicker in order to resist the greater heat of their fires.

Antonil (p. 141) assigned to each of his four teaches a specific purpose, and stated that the syrup passed through each in turn, although he did not specify their purposes clearly. Concerning at least the last of the four, however, his description suggests the same function as does the name of one of the teaches used in Mexico, as it was listed in an inventory of 1730 of Chiconcuac, a plantation very near Atlacomulco (T1969 E1). This last teach was called in Brazil the taxa de bater, at Chiconcuac the tacha de batir, and Antonil (p. 141) described the operation conducted in it as one of stirring with a bateideira, similar to a skimmer but without holes; part of the mixing consisted of lifting the filled stirrer as high as possible above the teach, after which the viscous liquid was dropped back into it.

Striking. In the last and usually the smallest vessel, the striking teache, the temperature of the syrup reached its highest point, and heating continued until it was decided that the grains would form upon cooling. Many rules existed to determine when this point had been reached. One English commentator remarked, "The Negro-Boilers have no Rule at all, and guess by the Appearance of the Liquor; and indeed it is wonderful, what Long Experience will do" (Baker, p. 34). It was not until nearly the end of the eighteenth century that Dutrône, with experience in sugarmaking in Saint Domingue, suggested the use of an hydrometer to analyze the density of the syrup (Deerr, Vol. 2, p. 582). The striking point was obviously of great importance, but before about 1780 it could not be objectively determined.

Ruiz de Velasco (1937, p. 249) described two methods of determining the striking point. In the first but less commonly used method, a skimmer or other ladle was used to take a little of the meladura from the striking teache (in the late nineteenth century called the plana), and it was whirled a few times on the ladle about the operator; the meladura, still hot, would form a film over the ladle if the striking point had been reached, and then fall off in fibers that would shatter. More commonly, the striking point or punto de azúcar was determined by taking a drop of the meladura on the tip of the thumb, then putting the tip of the index finger on the drop; when the latter reached hand temperature, the two fingers were drawn rapidly apart to form a thread that would remain intact if the meladura were at the striking point, or break near the thumb if it were not. This method was identical with the one commonly used in the British colonies. Ruiz de Velasco added that the breaking of the thread was characteristically accompanied by a dry cracking noise, and that pouring some of the meladura from a certain height back into the striking teache would result in a noise similar to that of rustling taffeta if the striking point had been passed.

When the striking point had been reached, the mass was transferred to a copper cooler, which in the Atlacomulco inventories of the 1680's was about twice the size of the teaches, and probably was a relatively shallow vessel with a consequently large cooling surface.

To receive the mass after it had been awhile in the cooler, the forms were brought from the adjacent room containing the tanks and racks wherein they had been prepared for use. New forms and pots had to be tempered for two or three days in a tank containing water with cane juice, some molasses, and some skimmings sufficient to cause fermentation that would seal the pores and make it easier to remove the loaf when purging was finished. Following this soaking, they were put into fresh water for about fifteen hours before they were used, to eliminate the odor of fermentation and probably also to end the process.

After cooling, the mass was transferred to forms similar to those shown resting on a rack (tendal in the Atlacomulco inventories) in the lower right-hand corner of Figure 13. The shoe-like scoop to the left of the rack was called in French bec du corbin (Labat, Vol. 3, pp. 275–76) and was used exclusively to transfer the mass from the cooler to the device lettered K and then to the forms on the rack. Its shape was intended to facilitate pouring, and it was filled by the instrument called remillon or cueilleur in French; a device used for the same purpose at Atlacomulco was called repartidera (reparteideira in Antonil, p. 143).

The filling of the forms was accomplished by a series of steps concerning which the Atlacomulco records contain no hint, but which was regarded as extremely important by both Labat and Antonil. Although their accounts differ in detail, they are in general agreement concerning the fact that the form was not to be filled in one pouring, but rather in two (Labat, Vol. 3, pp. 329–32) or three (Antonil, p. 143). In each case the hole in the apex of the form was first stopped, preferably with a banana leaf or stalk, and the forms arranged in groups of 4 or 5; each form in the group (called a venda by Antonil, ibid.), received an equal amount of the first of Antonil's temper, which had been cooked less than the following two, and all three were finally mixed well in the form. Labat gave no reason for the multiple fillings, but Antonil stated (p. 144) that unless this method were employed, using only temper that was perfectly cooked, the mass would be impenetrable to water, hence incapable of purging.

Authorities differed concerning the timing of the stirring of the mass in the form. Antonil recommended rather frequent stirring, whereas Labat (Vol. 3, pp. 329–32) waited for more than a quarter-hour until a crust had formed to a thickness that indicated stirring would be advantageous; probably the stirring shown in the lower right-hand corner of the figure represents this stage. Labat then marked the form to indicate that the first stirring had occurred, and returned in about half an hour for a last stirring. Any more than two stirrings would cause damage inasmuch as the sugar had begun to crystallize, but two were necessary to assist crystallization and also to distribute the grain equally within the form; furthermore, the stirring allowed grease to rise to the top and congeal, making it easy to remove (ibid.). About half a day later the forms were moved from the boiling-house, where the heat had kept the contents warm and the molasses somewhat thin, to the purging room.

Boiling-House Tools and Equipment. Only a very limited range of equipment was used in the boiling-house in addition to the kettles. Most inventories list wooden canals through which the cane juice ran from milling house to the receiver in the boiling-house, and the accounts show that pitch was bought to seal these canals. Skimming off froth and impurities after boiling had started was accomplished by means of an espumadera, most of which were made of tin with a long wooden handle. In addition to the skimmer, a long-handled ladle called a remillon was used. It is shown in Figure 13 together with the skimmer on the rack above the kettles where both were placed when not in use.

The Pottery

The Cortés mill always had facilities, called the casa (de obraxe) de formas, where pottery forms were made. Iron or wooden molds and one or two wheels were used for this purpose, and the vessels were baked in an oven (L269 E40). Indian or mestizo formeros were hired to make the forms, roof tiles, bricks, and whatever other rough baked-clay products were needed at the mill. In 1541, at about the same time that nearly 4,000 forms were bought in Texcoco for Axomulco, with more paid for their transport by Indians than the forms themselves cost, a Christian carpenter was hired to make a potter's wheel for the mill. In the 1540's and 1550's at least two Spaniards were hired to make roof tiles and forms and to instruct Indians in the art at Tlaltenango (L257 E14; L228 E3). In subsequent years many Indian and mestizo potters were hired to make the very large numbers needed; in the inventory of 1589, for example, more than 6,000 forms were on hand, adequate for about half the number of loaves of sugar produced annually at that time.

The substantial size of the pottery is shown by the entry in the inventory of 1760, where it was described as 85 by 22 feet, with an 8-foot ceiling. This was a much larger room than any shown in the plan of 1824, but at least part of the pottery in 1824 was like the description in the inventory of 1786, which listed two tanks east of the boiling-house used to cure and renovate forms, both in good condition. In the 1799 inventory a new galería for making forms was listed as under construction, and this may have been the "new" galería used in 1818.

Purging House

The shape of the purging house at Atlacomulco conformed to the recommendations found in most manuals: in the inventories of the 1680's, for example, it consisted of two large rooms, much longer than wide. The long narrow shape allowed the racks (barbacoas) on which the forms rested to be narrow enough so that it was not difficult to reach forms at the back of the rack.

The first important step in the purging house was to remove the stopper from the apex of the form, after which it was necessary to bore into the loaf through the hole at the apex in order to open a drainage way to a depth of several inches within the inverted cone. This operation was performed by using a bradawl ("furador" in Spanish and Portuguese). The forms were then put on racks, and the molasses was allowed to drip into canals that led to storage tanks.

The next major step, claying, required that the form be faced — that is, the fairly dry sugar at the upper end of the form was scraped to a shallow depth, then smoothed and tamped with a mallet in order to pack it, thus making sure that the water would not run too quickly through the loaf, and also to level the surface, so that the downward penetration of the front of water would proceed evenly. The prepared clay, moistened to the proper consistency, was then smeared over the prepared face (cara) of the loaf to a depth of two fingers (Antonil, p. 155). Four to six days later, according to Ruiz de Velasco (p. 254), it could be removed, and the face prepared for a second application, if necessary.

The second claying was not mandatory because, as Labat remarked (Vol. 3, p. 350), a second or even third claying — in fact, all the clay in the world — could not make good sugar out of bad. Ruiz de Velasco (p. 254) concurred, adding that some sugar after a single claying might emerge perfectly purged and whitened, and that more claying could only diminish the amount of sugar in the loaf. The sugar remained to drain for about a week after the second clay had been removed (ibid., p. 158), after which it was set to dry in the sun.

Keeping accounts of production was the responsibility of the purger; a sample page of a summary account of production — the daybooks do not exist — submitted to the Estate controller as part of an annual account of the plantation is shown in Figure 14. The production in this case ran from 15 June to 19 July 1615, in which period the mill processed 8 tareas of cane during as many days. The column on the left gives the number of cartloads per tarea, ranging from 18 to 20; the next column shows the numbers of arrobas (of 25 pounds each) of molasses produced, an unusual feature; and the column on the right, the number of loaves of sugar produced. At the top of each column is the total produced in the zafra to that point. The written material repeats the data in the columns, lists the date, and may include information about anything noteworthy or unusual; the first entry, for example, may be translated as follows: "On the fifteenth of June of the year 1615 there was milled another tarea of 20 arrobas, I mean cartloads, which produced 104 loaves and 10 arrobas of molasses."

Drying and Packing the Loaves

The earliest mention of sun-drying at Tlaltenango dates from 1543, when there was described "azucar blanco bien seco y asoleado" (L257 E12). The sun-drying platform was on the roof, with a storage room for sugar adjacent, and the first mention of an asoleadero in an inventory dates from 1655, when there was a small one over the much larger purgery. By at least 1786 the area of the drying floor was three times as large, and had besides a wall about 19 feet high and 3 feet thick. Although some authorities recommended using an oven with a capacity of five hundred or more forms, ovens were never used at the Cortés mill.

Once the loaf was taken from the form, all the poorly purged sugar was removed with a knife, an operation called mascavar (Antonil, p. 159) that gave the name mascabado to this grade of sugar. The foot of the form was separated from the rest to be returned to the purging house for further drainage (ibid.), but this may not have been necessary in Morelos, where the forms were much smaller than those used elsewhere.

The loaves were then wrapped in paper, first mentioned in 1543 (L257 E12), when the mayordomo of the Estate ordered the mayordomo of the plantation to wrap and weigh it before shipping. Balances with weights were used for weighing, and every inventory lists several of these. Upon arrival in Mexico City at the Estate store, the sugar was reweighed, and the head muleteer given a receipt that he took to the mill.

Grades of Sugar Produced

Although clayed sugar and molasses formed by far the greatest part of the production at the Cortés mill, small amounts of other kinds of sugars were made, as well as various conserves and sweets. In addition to the large amounts of clayed sugar produced directly from cane juice, the mill produced clayed sugar by boiling molasses: azúcar de espumas was made from the miel de furos, and the kind of sugar called panelas from the miel de caras. Often it was impossible to distinguish the latter from white, clayed sugar made directly from cane juice. Another type was called azúcar de respumas, but no meaning of this term is given in the Cortés documents. In the sixteenth century, the three sorts called respumas, espumas, and panelas were called as a group azúcares bajas, and in 1585 it was proposed to call all of them espumas in order to reduce the confusion surrounding the use of three names (L262 E2 D6). In the account of 1585 such a procedure was followed, leaving a simple pattern of two kinds of sugar made, white and espumas.

In the Atlacomulco accounts of 1702–9 appeared the first warning that the classification of sugar was to become more elaborate as the century wore on, with the first mention of the sugar called entreverada, a grade of clayed sugar that came from the middle of the loaf (Ruiz de Velasco, 1937, p. 253). The same document referred to a type called prieta, also the earliest mention of its kind, that was near the top of the loaf but at its circumference; Ruiz de Velasco showed a prieta primera and a prieta segunda in concentric arrangement, with the former nearest the center (ibid.). The full flowering of the classification may have occurred later in the century, when there had emerged terms based not only on the whiteness of the sugar and on the part of the loaf from which it was derived, but also on its condition: whether the loaf was whole (pilón), the sugar in lumps (pedacería), or in loose grains (polvo). Table 13 (p. 130), showing the price in reales per arroba, indicates the range of types of sugar sold from November 1779 to December 1781 (L232 E3).

In 1824 and 1825 the store records became rather simpler in respect of the number of kinds of sugar offered, since there were only white, entreverada blanca, entreverada corriente, and prieta. Of all the sugar received in 1825, 86 per cent by weight consisted of entreverada corriente and prieta, 12 per cent was entreverada blanca, and only 2 per cent was white. The fragmentary records of weekly production dating from 1905 show that in the eighteenth week of milling four color types — entreverada, corriente, mediana, and prieta — were produced, as well as a small amount of pedacería. Slightly more than half the total number of loaves consisted of mediana. None of these terms was in vogue in the sixteenth century, nor probably during much of the seventeenth, although substantially the same kinds of sugar were produced throughout the history of the plantation in substantially the same way.

Alamán (Documentos, Vol. 4, p. 532) explained to the Marqués that no mill in Morelos made refined sugar, which was "cosa de muy corto consumo," requiring apparatus and operations that differed greatly from those required for claying. Even the type of clayed sugar called blanca was more difficult to sell and offered less profit than the type called corriente. Thus it appears that the proportionately very great production of entreverada corriente and prieta in 1825, or of mediana in 1905, was reasonable in the light of the demand in Mexico.

Sugar Quality. Nearly every book or pamphlet ever written on the colonial sugar industry contains a reference to the poor quality of colonial clayed sugar. Much of the time the quality of the sugar made at the Cortés mill was not very high; an early and rather lengthy complaint (L208 E7), described above, made in 1586 by the seller of Tlaltenango sugar, Carlos Pérez, outlines the characteristics of low-quality sugar at that time. He had received in the preceding five years about 13,000 arrobas of white sugar, of which about one-third, although called white, was very dark and "muy ruin"; he had been storing in the warehouse some 3,000 arrobas that he had been unable to sell for more than three years, during which at least 10 per cent had to be written off. Nearly all of Pérez' witnesses agreed that a loss in weight at a warehouse of 6 or 7 per cent between the time of receipt and sale was not excessive, which means that the sugar commonly contained considerable amounts of molasses and water.

The accounts of other years describing conditions at the mill itself contain sufficient references to syrup that would not strike, or to sugar that was difficult to dry, to establish the fact that success was far from automatic in sugarmaking. Even if the sugar turned out well, it deteriorated in storage when the air was warm and moist (T1965; 1741), probably owing to the growth of yeast and fermentation. Or the cane might be poor, as in 1822 (L219 E2), when it was described as bad, past its prime, touched by frost, and damaged by rats; with these characteristics, only dark (prieta) sugar could be expected, so the mayordomo decided to make molasses instead. In addition to difficulties and uncertainties experienced at the mill itself, there were the further hazards of the trip on muleback to Mexico City, during which the loaves might become broken (*quebrado, en pedacería, polvo*) or increase in moisture content from humidity — all of which operated to depress their value and keeping qualities.

Production of Molasses

Ruiz de Velasco emphasized that mills run according to what he called the Cortesian methods produced primarily molasses, making its disposal a matter of some importance. It could not be carried far because of its low value in relation to its bulk, as well as the difficulty of producing tight casks, so most of it was sold locally. Although some was used to supplement the rations of the plantation workers, particularly the Negro slaves, and some fed to livestock, most was sold and some thrown away. Wherever molasses was produced, the revenue obtained from its sale was expected or hoped to be enough to cover production costs of both sugar and molasses, leaving the entire sales value of sugar as profit. Alamán (*Documentos*, Vol. 4, p. 532) wrote in 1851 that the sale of molasses generally covered all costs, and that it was reasonable to expect an annual revenue from it of $15,000 to $20,000, at a time when about $500 in cash was needed weekly to run the plantation.

The molasses must have been intended largely for the manufacture of alcoholic beverages, particularly aguardiente, even though the practice was frowned upon by the authorities throughout the colonial period. A viceregal edict (L79, Hojas Sueltas) of 10 June 1608 prohibited the sale of molasses in cities and pueblos to Indians for any purpose, including eating. It was possible to make aguardiente in the eighteenth century under license, and it has to be assumed that much of the molasses made at Atlacomulco was sold to distillers in Cuernavaca for this purpose, since most went to a few large buyers; some small amounts were sold to Indians.

The purgador was supposed to account for molasses sales and production, but his daybooks no longer exist. Only in a few years, discussed in detail below, was total molasses production listed with total production of sugar in the annual accounts. For some years, only the total income from molasses sales is available, but unit prices were hardly ever stated. However, from the period 1585–1625 enough data remain to show that the amount produced depended in part on varying circumstances of weather, seasonal changes in weather, age of the cane, and fires, as well as to suggest that before Independence an effort was made to minimize the amount of molasses produced.

Kinds of Molasses Produced. Four kinds of molasses were produced: miel de furo, miel de caras, miel de barros, and miel de tareas. Miel de furo was defined in 1619 and 1621 (L269 E16; L260 E22) as the first molasses to drain from the forms. The amount of miel de furo obtained was greater in every year than the volume and weight of the two kinds of molasses obtained subsequently by claying, the miel de caras and miel de barros (also called miel de segundo barro). The latter two were usually grouped together in the accounts, making it impossible to determine the amounts produced of each. The miel de caras resulted from the first claying, the miel de barros from the second. Both were used to make panelas, so I have omitted from the summary (Table 16, p. 131) the small amount of panelas made in order not to count these kinds of molasses twice.

All the rest of the molasses produced at Atlacomulco was called miel de tareas, made intentionally and directly from cane juice. In the accounts of 1611 and 1612 the miel de tareas was recorded separately from molasses drained by purging, and the account of molasses also explains why molasses was made from all the cane in the tareas. For example, all the cane yielding only molasses in 1611 came from fields where the cane had burned, apparently by accident. In 1612, by contrast, all the cane was either first or second ratoon, and was described as ruin or seca. In general, the yields of molasses from the damaged cane of 1612 were much lower than the yields of the burned cane of 1611.

Seasonal and Interannual Variations in the Production of Molasses. For some of the years between 1610 and 1625 the annual accounts describe seasonal variations in the production of molasses and sugar. Table 16 shows dry season production (11 January to 1 May) and wet season production (11 June to 29 July) for some years. The relationship in the dry season between loaves produced and pounds of molasses drained seems fairly direct; by contrast, in the wet season widely varying amounts of sugar might be produced from the same number of loaves. In general, much more molasses was produced per loaf of sugar in the wet than in the dry season, a fact also expressed in the large differences in the average weights of loaves for the two seasons: 10.7 pounds per loaf in the dry, but only 7.8 pounds per loaf in the wet, indicating considerable drainage from loaves containing originally a relatively high proportion of molasses. The total weight of molasses and sugar per loaf in the two seasons also differed markedly, being 12.2 pounds in the dry season and 16.9 pounds in the wet.

When the annual production in the early seventeenth century of each of the two kinds of molasses, the miel de furos and that obtained by claying, is expressed as the ratio of weight of sugar produced per cart to the weight of each of the two kinds of molasses, and the two ratios are plotted, the values fall on a straight line. I have estimated the relationship to be expressed by the function $Y = 1.94X - 6.4$, where X is the ratio of weight of sugar to weight of molasses produced by claying. In sharp contrast to this marked linear relationship is the lack of any relation between weight of sugar and all kinds of molasses, including miel de tareas, produced per cartload of cane. These results might be expected in a case where it was regarded as desirable to minimize molasses production, but lack of control over the sugar-making process and the quality of cane made the goal impossible.

Table 17 (p. 131), summarizing data concerning number of loaves and weight of sugar produced per cartload of cane from 1587 to 1625, suggests that when sugar was made with more or less success, the relationship between the cartloads of cane and the sugar produced was fairly steady. Thus, variations in annual molasses production derived largely from the state of the cane and were expressed largely in variations in the production of miel de tareas. In two periods for which fairly detailed information is available, mostly in the early seventeenth and early nineteenth centuries, there seems no doubt that the aims of production differed between the two periods, since in the earliest the ratio of weight of molasses produced to weight of sugar was fairly low (mostly less than 1:1), whereas in the later period it was higher, in every case more than 1:1 and in half the cases only slightly less than or slightly more than 2:1. Furthermore, it should be noted that in the first two decades of the seventeenth century large stocks of molasses were on hand at the end of the year, suggesting that it was difficult to sell and that, given a choice or the ability, the amount of molasses produced would have been even lower than it was in relation to the amount of sugar produced.

Changes in Mill and Boiling-House

Eighteenth-Century Changes. Since Atlacomulco was leased nearly continuously between 1721 and 1763, it is difficult to fix precisely the dates of some important changes in the sizes of parts of the mill that apparently occurred during these 43 years. It seems likely that most of the improvements were undertaken by lessees with the aim of increasing output and in response to low prices.

Information in inventories and annual accounts shows that there was little change in such important features as the average weight of kettles and of bearings supporting the waterwheel until the late seventeenth century and early eighteenth century (Fig. 15 and Table 18, p. 131). The average weight of kettles nearly doubled between 1721 and 1746, and changes in the weights of bearings amounted to nearly 20 pounds sometime between the inventories of 1558 and 1702, after which they remained about the same until 1799. Some of the change is obscured by the fact that in the early eighteenth century, until the middle 1740's, the mill was converted from water to animal drive. Possibly the weight of the cylinder sleeves of the mill and the bearings was less during that time owing to the weaker power source; heavier bearings did not appear until the inventory of 1763, when the reconversion to water power had occurred and some of the bearings weighed 85 pounds. Although the evidence is inconclusive, it seems likely that by the end of the third decade of the eighteenth century information available to the managers of the Estate and the lessees of the plantation suggested that significant increases in the weights of major parts of the mill and boiling-house were both desirable and technically feasible.

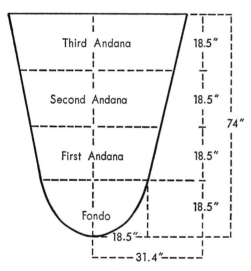

Figure 15. Construction of kettles used at the Cortés mill

The increase in the size of kettles continued in the mid-eighteenth century, but was not matched by significant changes in the weights of sleeves of the milling cylinders, and little change occurred in the weights of the main bearings of the waterwheel. The numbers of kettles stabilized at ten for the rest of the century; although fewer than those in use in 1585, they were much heavier, and probably also much larger, than they had been when the mill was one of the largest in the Indies.

Another important change occurred with the elimination of the bagasse presses in the late 1750's. In 1798 two waterwheels were installed to run two sets of rollers, and in 1760 a foundry was established for casting the much heavier kettles, bearings, and sleeves.

Nineteenth-Century Changes. No important changes in boiling-house equipment seem to have occurred after the 1750's until 1884, by which time the number of kettles had been increased to fourteen arranged in pairs (mancuernas); their average size appears to have remained about the same. However, major changes had occurred in milling equipment. The inventory of 1831 shows that sometime in the early nineteenth century the weight of bearings nearly doubled, from 60 or 80 pounds to 125 pounds by 1831, and the sleeves of the milling rollers became enormously heavier, increasing more than half a ton in weight between 1799 and 1831, from 1,550 pounds to 2,650; at the latter date, in addition, they were described as cast.

After the American invasion and occupation of central Mexico, conditions in the sugar industry began to improve, and demand created by the California gold rush sent the price of sugar soaring by early 1850, with the result that improvement of the mill became one of Lucas Alamán's major preoccupations. An aguardiente factory was added in 1852, the change to horizontal rollers was made in 1854, and there was discussion concerning discontinuance of carriage of cane by mules in favor of carts, as well as the addition of some shallow and flatbottomed (plano) kettles to the battery.

However, the changes in scale of operation and in equipment appear minor in contrast with two major changes in sugar manufacturing that did not reach Morelos until after the Revolution: the use of steam in processing and milling, and associated conversion from the traditional and expensive production by relatively small plantations to the central system. Even by 1905, the last year from which there survive letters from the mayordomo to the administrator of the Estate, these topics were merely under discussion, although they were broached with some urgency because of a poor harvest.

In the long time between the establishment of the mill in 1535 and its abandonment early in this century, it appears that very few changes were made in equipment or processes except possibly the introduction of the three-roller vertical mill and the substitution of materials in milling — iron for wood — that not only ensured more dependable operation but also permitted the use of large equipment and increases in capacity. The mid-nineteenth-century changes were late as far as their adoption elsewhere was concerned, but they did increase the capacity of the mill and output per worker. It is therefore correct to say, as did Ruiz de Velasco in his description of the industry in the late nineteenth and early twentieth centuries, that until the Revolution the form — but not the dimensions — of much of the equipment of the mills of Morelos had changed but little since the days of Cortés, but it will be shown below that it would be incorrect to infer that they had remained at the same level of performance.

VII

Livestock and Supplies

INASMUCH as the impression is widespread that sugar plantations aimed at self-sufficiency because, in part, of an assumed isolation, it may be worthwhile at the outset to dispel this notion by stating flatly that the Cortés plantation produced nothing but clayed sugar, some confections, and the inevitable molasses, as well as some dried but not tanned cowhides and sheep pelts that were the by-products of animal mortality and the custom of eating meat. Apart from these products, the only other item steadily produced was a small amount of lye necessary for sugar manufacture and perhaps for other uses.

The best time for examining the "self-sufficiency" of the plantation is the sixteenth century, when the industry was still relatively new in Mexico and it might be assumed either that transportation was at its most difficult or sources of supply of goods and livestock had not yet become fully developed. Almost every annual account from the 1580's shows that practically all livestock and every item consumed or used on the plantation was bought, including even leather goods — straps, reins, cinches, leather chests, possibly shoes for Negroes — and such items as small nails and metal parts for carts and mill, at a time when a smithy was in full operation and adequately equipped to supply such items. A large range of wood products—beams, laths, shingles, handles (perhaps in the rough) for tools, yokes, planks, barrel staves, charcoal, pieces of wood of specified sizes

for building, and vast quantities of firewood — were *purchased*, not obtained by the labor of plantation slaves or Indians hired for the purpose of supplying them under the supervision of plantation personnel. One can see from this list of parts, as from the list of metal products bought, that the primary purpose of the Spanish tradesmen and skilled Negro slaves and Indians was to assemble and repair needed articles rather than to manufacture them from raw materials obtained locally.

Although the plantation produced some corn in the eighteenth and nineteenth centuries, at this earlier date most of the corn consumed was assigned to the plantation from Estate tributes, and even some of this had to be brought from as far away as Toluca. The list of purchased food other than fresh meat — even much of the latter in the early seventeenth century was bought from the slaughterhouse in Cuernavaca — must be a nearly complete list of sixteenth-century diets: wine, vinegar, fresh and dried fish, shrimp, hams, raisins, almonds, lentils, garbanzas, beans, cacao, vanilla, honey for confections, limes, and even chicken and eggs (the latter for the sugar-boiling process and the infirmary) were bought off the plantation and carried to it. Even zacate (a grass) was purchased cut for the horses, and some of it was bought to make hay. Tallow, used in large quantities on bearing surfaces, and lard had to be bought, even though some must have been available from animals slaughtered on the plantation. Soap was bought, although both lye and tallow were available on the plantation. Wax was purchased to make candles, but on the other hand finished candles were also bought. Cloth for strainers, clothes, pitch and oakum, and large quantities of salt for eating and for curing hides had to be purchased as well; in 1584 alone, 3 tons of salt (sal tierra) were bought for the latter purpose. Building, as well as building repairs, required the bringing of a never-ending supply of lime (cal) to the plantation, supplied by purchase from local Indians. Even cane for planting and for milling was bought frequently in the sixteenth century, and the plantation never became self-sufficient in Negro slaves and draft animals through natural increase; the ranch Mazatepec and local entrepreneurs were relied on to supply the latter.

Perhaps more important, there never was any suggestion from Estate officials that it should become self-sufficient in any of these lines, and the only suggestion that plantation labor might be used to supply any staple occurred from time to time in the case of firewood, but this goal was never achieved (Appendix E, p. 115). It is necessary to assume that from the beginning of the sugar industry in New Spain, entrepreneurs operated within a system characterized by a high degree of specialization and the ability to supply through purchase the range of items required by its parts.

Livestock

For about thirty years after the founding of the plantation in 1535, the contribution of Indians to transport was extremely important, but after depopulation it became necessary to increase the number of oxen at the plantation and to rely on mule trains rather than Indians to carry sugar to Mexico City. Apart from oxen and mules, other livestock were of much less importance: horses were used only as mounts and to produce mules, but not for draft, nor carrying cane and sugar, nor powering trapiches; sheep were bought for rations and their pelts.

OXEN

Numbers of Oxen. Table 19 (p. 131) indicates that the number of oxen at the plantation fluctuated widely from one inventory to the next. Before 1600, numbers seem to have been kept near 150, but in 1585 twice as many were present. The low number of 50 held in 1600 may be explained as a result of the disorganization of both the Estate and the plantation at that time. The low number of 1634 expresses one of the difficulties introduced by leasing, the maintenance of stock numbers; in this case, replacement of the missing stock was made quickly by the heirs of Andrés Arias Tenorio. From the late seventeenth century until the inventory of 1812, numbers of oxen were maintained above 500, with a peak of 658 in 1807 that had been equalled only by the previous peak of 657 in 1679. The difficulty of maintaining the oxen population after the beginning of

the Revolution is well illustrated by the drop in numbers that occurred between 1812 and 1818, and even after Independence their numbers continued to decrease. This decline does not necessarily signify decreased dependence on oxen, but perhaps instead the rise of a group of peasants who owned their own oxen and hired them out for use at the plantation.

By the end of the sixteenth century Tlaltenango was able to produce between 50 and 75 tons of clayed sugar annually, and it is worthwhile to know how many oxen were regarded as necessary to sustain that rate of production. The average of the eleven figures (Table 20, p. 132) representing inventory stocks at different times in the early seventeenth century is approximately 350, but six of the figures exceed 400, and some number greater than 400 should be taken as the goal. This conclusion is supported by a statement made by Governor de Molina (L258 E9), when he was attempting to rehabilitate the plantation in 1614, that a mill of its size required between 400 and 500 oxen, according to experienced persons. As Table 19 shows, later in the century their numbers were indeed maintained above 500.

Mortality and Replacement Rates. The only source of information concerning the replacement rate of oxen at Tlaltenango is the annual accounts for the years 1609–25, which contain sufficient detail (except for 1618) to yield an average annual mortality rate for the entire period that is shown in Table 20. Those that were culled were eaten, and I assume that the figures in the column headed "Lost" are residuals obtained by comparison of losses through death and culling with other information about their numbers.

The range of annual mortality rates suggests an important element of interannual variability that added to the unpredictability of the outcome of the efforts of any year, since it was high, between 5 and 33 per cent; in the latter case one might suspect culling as the result of a large purchase of 358 oxen in the same time period, except that the account states that the oxen died. Perhaps the range in mortality rates revealed by these data is a fairly adequate description of the possible range, extending from a low annual death rate of one in twenty to a rate of loss so high as to cause the death of one-third of the herd. In the latter case, it should be kept in mind that the high rate of loss is necessarily an average for a period of nearly three years, and if the losses had been concentrated in a single year of the three they would have greatly exceeded 30 per cent. The average annual rate of loss for the approximately fifteen years is 13 per cent, giving an average working life of an ox of about 4 years, assuming that work began at about 4 years.

Causes of Death. That epidemics among oxen did occur in Morelos is stated by the administrator Robles (1905; L240 E15), who wrote that for some months no oxen at Atlacomulco had fallen sick in spite of a local epidemic, and that a second vaccination had been given the herd in order to prevent the "mal." Except for this brief notice, there is no information concerning the nature of the illnesses that affected oxen, although in numerous accounts losses were attributed to sickness. Perhaps another epidemic occurred in 1799, since the inventory of that year listed 515 oxen, with the information that more were needed because 93 (16 per cent) had died in the year.

A statement (L268 E3B) from the early seventeenth century identifies other sources of loss: drought, failure of pasture, lack of shade, a wide scattering of animals that caused them to be overlooked at roundup time, and attacks on livestock by Indians. Although two or three mounted boyeros attended the herd, they were not expected to be able to keep the animals together because the grazing was so sparse. Straying made damage to Indians' crops inevitable, and they responded by killing or maiming stock. Nevertheless, this system of grazing, with its attendant losses, was cheaper than fencing, and it was the practice of the plantation mayordomo to pay damages to Indians without recourse to the courts in cases involving stray oxen, cattle, and horses. Nearly every annual account lists expenses due to "daños de bueyes," hardly ever exceeding $75.

Work Performed by Oxen. Oxen performed only three tasks: hauling cane carts, pulling plows, and turning the bagasse presses; oxen trained to one task were apparently not used for any other.

Until the late eighteenth century, most oxen were used to pull cane carts. Both the numbers of carts and the numbers of oxen to pull them remained fairly high (7 to 10 carts and 300 or more oxen) in the latter part of the sixteenth and throughout the seventeenth century, but probably peak numbers had occurred earlier: the Relación of 1556 (L267 E26) states that there were 50 carts at Tlaltenango and 4 Negroes to manufacture and maintain them. The latest inventory with large numbers of cart oxen dates from 1688 (333) and their numbers never again reached that level. Although very few carts were kept in the eighteenth century, cart oxen

nevertheless were important until the latter part of the century. The difference in numbers of oxen used for this purpose in 1763 (199) and only 14 years later in 1777 (106) illustrates dramatically the sudden drop in their numbers upon their replacement by mules for carrying cane. In the succeeding two decades, the numbers kept to pull cane carts declined even more, to below 100, but there was a slight increase in their numbers after the beginning of the nineteenth century. In 1847, with 12 carts listed in the inventory, no cart oxen were listed; instead, the 245 oxen consisted of only two groups trained for special plowing tasks.

Each cart was pulled by a pair of oxen, and the ratio of cart oxen to carts was extremely high during the period in which both were important: there were often between 25 and 35 oxen per cart, and even if their numbers had been halved, the ratio would have remained high. The account of 1585 (L262 E2 D5), showing daily production of sugar as well as inputs in cartloads, indicates that no more than 26 carts of cane were brought to the mill in a year when only 26 carts were listed in the inventory. In the same year, the mill owned 270 cart oxen, giving a ratio of approximately 10 oxen to one cart, requiring their use once every five days if they were used in pairs; in fact, the mill did not operate every day, with the result that each ox may have worked one day in six. This frequency of use seems extremely low, but in fact it differs little from that implied by Avalle's (p. 3) figure of 8 oxen per cart in 1788. The frequency of use in plowing was somewhat greater, since there were 9 plows listed in the inventory of the same year and 37 plow oxen, a ratio of 1:4. Neither planting nor milling occurred in every month of the year, although the distribution was less concentrated in time than it later became, with the result that the oxen may have worked over a longer span of time in 1585 than in the eighteenth century, but the total work done by each ox cannot be regarded as great.

Before 1625, fewer than 50 oxen were kept for plowing, and their relative unimportance can be explained only as a result of the fact that ground was prepared for planting by hand, with heavy reliance on the large numbers of Indians available, and on Negro slaves. A sudden increase in the number of oxen used for plowing occurred between 1625 and 1654 (see Fig. 16, below), and their numbers doubled between 1654 and 1672. This feature of their numbers was matched, at least between 1635 and 1655, by a great increase (from 4 to 18) in the number of plowshares, which generally bore a close relation to the number of plows. However, the number of plows does not seem to have increased significantly again until sometime after the first two decades of the eighteenth century, when it doubled. Another doubling of the number of plows occurred sometime in the last quarter of the eighteenth century, matching an increase in the sugar output of the mill, but whereas the latter increase is reflected in the num-

ber of oxen used for plowing (100 in 1763, but 347 in 1777), the earlier eighteenth-century increase in the number of plows was unaccompanied by an increase in the number of oxen used for plowing.

Quality of Oxen. The only specifications I have found of the attributes desired in oxen are contained in a letter from a mayordomo (L30 V56 E1) who wrote on 17 August 1707 that he had received 72 novillos but that

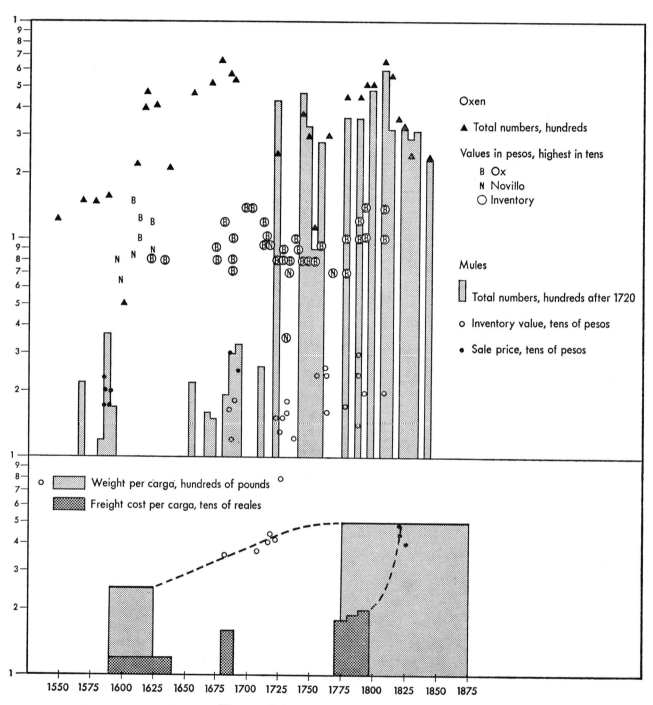

Figure 16. Prices and numbers of livestock

they were unsuitable for their purpose, being too small and young. He recommended that novillos sent to the plantation from Estate ranches should be four or more years old. From this statement it appears that the cattle grew slowly, as might be expected of a breed so small; their size is discussed later.

Management of Oxen. The author of *Instrucciones a los Hermanos . . .* (p. 112) recommended use of a movable cattle pen for various reasons, partly because of savings in time if the cattle were kept near the work, and partly to facilitate dunging. The recommendation is interesting because the practice was common in parts of the British West Indies at about the same time (1725–50), but there is no indication that dunging was ever done at Atlacomulco. However, a corral for oxen and another for mules near the mill were standard.

MULES AND HORSES

The sudden rise of mules to prominence on the plantation in the first part of the eighteenth century is so striking that it might be regarded as a technological innovation, were it not that mule-driven trapiches had been present in Morelos for at least a century: for example, the two trapiches at the hacienda Tlalcoalpanque (T1725 E1) were driven by 56 mules, and the ratio of mules to oxen there was about 1:1 in the early seventeenth century. The Atlacomulco inventories, on the other hand, indicate that until after the end of the seventeenth century no more than 50 mules were held by the plantation. Suddenly, in the inventory of 1721 there appear 435, perhaps a tenfold increase in their numbers, as well as a sudden increase in the number of mares, evidently intended to maintain the number of mules (Table 21, p. 132, and Fig. 16). Most of the increase can be accounted for by the abandonment of the traditional water power in favor of mule power to run two trapiches, as is evidenced by the number of mules (slightly more than 300) described in the inventory as trapicheros. At the same time, the number of mules used by employees in place of saddle horses, as well as the suddenly large numbers (approximately 50 and 100) described as cañeros in the inventories of 1721 and 1741, indicate that they had come into favor at Atlacomulco for a wider range of uses than formerly. After the abandonment of the mule-driven trapiches and restoration of water power, large numbers of mules — well over 300 — were nevertheless retained. In fact, peak holdings occurred in 1799, when almost 500 were present. From this time on, their numbers declined, but

holdings remained larger than they had ever been before 1700.

There is no reason given for the great change in the reliance upon mules in any documents, so it is necessary to attempt inferences from whatever data seem to have a bearing on this rather puzzling change. It should be remembered that mules had long offered the only means of transporting sugar to the capital, and the plantation had never depended on its own train or requa to perform all transport tasks in the sixteenth and seventeenth centuries, since the size of its requa (24 in 1566, 37 in 1585, and none from the 1630's to the 1680's) was inadequate. Instead, there was long-continuing dependence on rental of the requas of persons in the Cuernavaca region, and many other plantations must have met their needs in the same way. For transport, therefore, the use of mules was well-established before the suddenly increased dependence on them early in the eighteenth century.

Furthermore, the Estate had long been engaged in breeding mules at Tehuantepec, whence they were driven to Mazatepec and sold to owners of requas and livestock dealers. The resources of Tehuantepec for producing both horses and mules were certainly adequate by 1556, the date when a short description of one ranch there showed it had 10 jacks and 400 mares, and another was stocked with both also (L267 E26). Their progeny were driven to Mazatepec, whose accounts (L262 E3) for the years 1587–89 show that in the first year there were only 60 mules at the ranch, of which 18 were lost within the year, having "fled toward Tehuantepec"; in 1588, however, 271 arrived, together with 52 colts. Some of the newly arrived mules were sold before the end of the account period, the leading purchaser being Diego de Benavides, a livestock dealer of Cuernavaca on whose requa the plantation depended heavily at this time. He purchased 33 of the mules, 11 were assigned to Tlaltenango, and 3 were given to various persons in Mexico City. Interestingly, the only purchasers of horses were Indians, who bought 34 of the 52 colts. In 1590, 136 mules were brought from Tehuantepec to Mazatepec, with 22 mares. The Governor de Quintana Dueñas released 232 of the mules and 91 colts to Benavides, who had thus purchased in three lots and three years 580 mules.

Some mules were also produced at the broodmare ranch called Tlaltizapán, adjacent to the pueblo of the same name in Morelos. In the inventory of 1549 this ranch had 107 mares, "de todos colores e gordas e crecidas, e de buenas personas." In addition, there were 59 fillies up to 18 months of age, 54 colts of the same ages, and 9 young mules, so it appears that Tlaltizapán was producing primarily horses of good quality, but a few mules as well. At Atlicaca in Morelos (*ibid.*, pp. 287–88) there was a pasture for colts, containing at the same time 61 unbroken colts of good quality, and 18 mules from two to four years old, also "de buenos cuerpos y personas."

From the ranches at Tehuantepec and near Cuernavaca itself, then, it appears that even by the mid-sixteenth century it was possible for the Estate to produce annually several hundred mules, perhaps even mules of good quality, and that by the latter part of the century enough were delivered to Cuernavaca to maintain easily a stock of mules 400 or 500 strong at the plantation, but dependence on mules remained slight in spite of their availability.

Lamb has reviewed the historical geography of the mule in the southern United States, describing the rationale developed there to justify preference for them over other types of draft animals, largely in terms of their hardiness (p. 82); this opinion is found also in the work of Fanny Calderón de la Barca (p. 354), whose pages are filled with references to mules. Lamb's conclusion, however, is that "no irrefutable proof has been offered to indicate the superiority of the mule," but that the preference is a culturally conditioned "personal preference" (p. 83), and, indeed, Spanish mules and the preference of the Spanish for mules have long been famous.

Ruiz de Velasco, whose works are informative concerning many questions, offered some suggestions that may help to explain the sudden prominence of mules at Atlacomulco. He stated (1937, p. 144) that the population of Morelos was a "raza muy característica," formed through the interbreeding of Spanish, Indians, and Negroes, and called gente de razón; this group was in his mind sharply distinguished by attitudes toward work from the Indians of the same area. He emphasized the difference (*ibid.*, pp. 186, 190) in terms of their contrasting preferences for mules and oxen, stating that the Indians preferred oxen, but that the peón de razón always chose the faster mule. In part, the choice was based on two alternative wage plans available for workers in Morelos, which he also regarded as peculiar to the state: one could choose to be paid either by the day or by the task, and here again the choice of the two populations differed, with the Indians choosing the daily

basis, but the peón de razón preferring payment by the task. Once the choice was possible, there existed a reason why mules should continue to be important at the plantation through the eighteenth and into the nineteenth century after the need for large numbers to power the trapiche had disappeared, in order to satisfy the different preferences of the two social groups, one of which — the gente de razón — was becoming increasingly important in numbers. It is also important to note that the large numbers of mules were kept primarily for plowing rather than transport.

Horses. Table 21 (p. 132) shows the slight dependence placed on horses, in marked contrast to the high dependence on mules and oxen. Scarcely any horses were kept before 1600, but in 1655 a record number (121) was present. Even the relatively large number (113) held in 1811 included many colts. The horses were used when bringing stock from Mazatepec, as mounts for the oxherds, for producing mules, and as mounts for the mayordomos, field guards, and vaqueros. The vastly different livestock situation at the plantation in the late nineteenth century is pointed up by the inventory of 1884, which lists only 7 horses used as mounts. In spite of their relative unimportance at the plantation, large numbers were available at Mazatepec, where the late sixteenth- and early seventeenth-century accounts show upwards of 300 head pastured.

MANAGEMENT

Cost of Livestock. After about 1625, nearly all the available values of livestock occur in inventories, not only from Atlacomulco but also from other plantations in Morelos. These inventory values are consistent with one another in cases where they are available from different plantations at approximately the same time, and it is reasonable to assume that they approximate market or replacement values.

The values I have found are presented graphically in Figure 16, together with the total numbers of animals at the Cortés plantation. The oxen values are the most complete and show that, throughout the period covered by the data, the range of values at any date extended from about $8 to about $15 per head, depending on condition, age, and speciality, with those used at the bagasse press generally the most expensive, followed by cart oxen and plow oxen in that order. The price may not have changed much in the later nineteenth century, since the inventory of 1884 showed 169 "regular" oxen valued at $15 each. The inventory value of mules was generally about twice that of oxen, and three or more times that of saddle horses. The range of mule values at any time depended on the same circumstances as did the range of ox and horse prices. Horse values were generally low, and it appears that from 1600 to 1825 it was possible to obtain an adequate horse for $3 to $5.

It is difficult to establish trend with a series so incomplete, but there is a suggestion of low value of oxen from about 1725 to 1775, occurring simultaneously with a reduction in their numbers at Atlacomulco. A similar decline in the value of mules is also suggested in the early part of the same period, but a recovery in their price may have occurred earlier than the recovery in ox prices. In any case, it appears that at a time when the price of mules was lowest, their numbers increased greatly at Atlacomulco. As the lower part of the graph shows, this was also the time at which the average weight of the carga or standard mule-load of sugar was nearing its peak, and both these changes may have affected the rapid growth in the number of mules.

Costs of Transport to Mexico City. The history of changes in freight rates and weights of loads is summarized in Table 22 (p. 132). No information is available concerning transport charges before 1591, when 241 mule-loads were carried at a cost of $1-4-0 per mule for the distance of about 60 miles to Mexico City. This rate held until at least 1607, the last date at which the rate was presented in the annual account. By 1644, the date of the next available information, the rates varied from the same as that charged in 1607 to the lower figure of $1-0-0. The next available information, dating from 1681, shows that the cost had risen to $2-0-0, where it remained for approximately 30 years. The rise was compensated, however, by a rise of 100 pounds or more in the weight of the average carga of sugar, distinctly different from the 250 pounds, or ten arrobas, stated to be the average weight in 1607 (L268 E3), then equal to the weight of a carga of firewood (L268 E1). Some hides were also carried, with a carga defined as consisting of 8 hides, but no weights were given; the amount charged was the same. The next available information is from 1754, but since the number of loads was only 22, the price and weight per carga may not be representative. The cost of freight had risen again by 1773, but the weight of the load of sugar had reached the figure of approximately 500 pounds, including the trunks, that it was to retain, double the amount carried by a mule only 150 years before. A rise of one real per load occurred in both 1782 and 1790, a doubling of the freight rate occurred sometime in the revolutionary period, and after 1824 no information is available.

Lower rates were available on the return trip from the capital, amounting in the 1590's and in the first decade of the seventeenth century to only $0-6-0 (L262 E5 D7; L263 E15; L268 E3), because so much less freight went from Mexico City to Cuernavaca. It appears that the number of cargas of sugar and hides sent to the capital at that time ran about eight or ten times the number of loads of supplies — principally iron and clothing — sent to Cuernavaca. In 1600, for example, 392 cargas were carried to Mexico City, whereas only 45 loads were carried on the return trips (L263 E15). Probably this difference persisted throughout the history of the plantation.

There can be no doubt about the change in the weight of the carga, and at least part of the explanation involves changes in the weight of a loaf of sugar. The carga always consisted of 22 loaves of sugar, which ranged between 10 and 15 pounds in weight in the sixteenth and early seventeenth centuries, increased to above 15 in the late seventeenth and to somewhere near 25 pounds in the eighteenth century. Atlacomulco was not unique in respect of this change in the weight of a loaf of sugar, and the abundant documentation concerning the weights of loaves leaves no doubt that it occurred.

On the basis of the information I have, however, it is not possible to state whether the change in the size of the loaf was a response to changes in the freight rates, or to an improvement in the mules themselves.

In spite of the high costs of shipping sugar, requiring large cash outlays on the part of the Estate to residents of Cuernavaca, there was surprisingly little effort to remedy the situation by maintaining sufficiently large numbers of mules for this purpose at the plantation. Generally, the mules made the trip in groups of 20 or 40, thus requiring in the average year from 25 to 50 trips, depending on the number of mules in each — that is, a weekly or twice-weekly departure, with most hauling taking place in the dry season. Probably 150 mules would have been adequate to maintain a twice-weekly schedule, yet even when large numbers were kept at the plantation in the eighteenth century, freight costs still appear as amounts disbursed to requa owners.

Probably sometime in the eighteenth century another approach to the problem of ensuring a local supply of mules for hire was initiated. Mule owners, probably mestizos, were assisted by loans from the haciendas to purchase and maintain requas, which were then hired out to the haciendas. In 1814, at a meeting of the Junta of the Estate, such loans were discussed in connection with four applicants, each of whom had a few mules and wished to purchase more but lacked cash (L219 E1). The mayordomo of Atlacomulco, who had initiated discussion of the matter by presenting it in a letter, wrote that all four were good men, vecinos of haciendas in Cuernavaca, that it was customary for haciendas to make loans for such purposes, and that the Estate could rest assured that the loans would be repaid. In any case, he wrote, there was little choice in the matter, since without more mules the sugar would remain at the mill.

Efficiency of Animal Use at Atlacomulco. Table 23 (p. 133) offers a basis for judging efficiency of stock use at the plantation, by showing at different times its total production of sugar and molasses and total numbers of stock, in comparison with estimates of numbers necessary to produce various amounts of sugar in the British and French possessions. What emerges is the extreme disproportion between numbers of stock and total production at Atlacomulco; the only similarity is from the mid-eighteenth century, between Belgrove's figures and the Atlacomulco inventory of 1753–54, yet 1754 was an unusual year at Atlacomulco, with stock numbers far below normal. In the period 1718–21, when annual production of sugar and molasses was about 220 tons at Atlacomulco, there were about 800 head of livestock; later in the eighteenth century, between 900 and 1,000 head were kept at a time when it could produce between 350 and 400 tons of both annually. The efficient Saint Domingue plantation described by Avalle in the 1780's produced nearly the same amount of sugar and molasses with only 121 cattle and mules; even Phillips' example, not the most efficient producer in Jamaica, produced about the same weight as did Atlacomulco, but with only 220 mules and oxen. Thus, the data available from the late seventeenth to the late eighteenth century suggest that from five to eight or more times as many head of livestock were maintained at Atlacomulco to produce approximately the same amount of sugar as on French and English plantations.

Supplies

Supply and Costs of Wood. One of the major expenses of the plantation was incurred in buying firewood. Probably nearly all the wood was cut by Indians, with the Spanish owners of mules assuming the role of middlemen, although it appears that in the first two or three decades of operation of the mill the Indians both cut and delivered the wood, since at that time it was delivered by the braza rather than the carga.

Sugar mills were voracious consumers of fuel, as may be seen from the data on supplies between 1580 and 1625, when between 10,000 and 20,000 cargas were received annually. The carga weighed 250 pounds in the late sixteenth century, so between 1,250 and 2,500 tons of wood were consumed each year by the furnaces.

There is no close relation between the amount of wood consumed in a year and the amount of sugar produced; instead, the scatter is very great. Probably this scatter is not due to carryover of wood from one year to the next, because, although most inventories list a corral de leña and its contents, it never contained more than a month's supply. Still, it would be most interesting to discover the relation between wood and other fuel consumed and sugar produced, because of the possibility that changes in this relation might express gains in efficiency or increased use of bagasse to fire the furnaces. The data pose a problem of interpretation; for example, in the period 1580–1625, the supply fluctuated between 10,000 and 20,000 cargas per year, and in the period 1770–1830 it fluctuated much more widely, from a little more than 4,000 to approximately 33,000 cargas per year. Although five of the eleven annual values fell between 12,000 and 16,000 cargas in the latter period, much more sugar was produced by the mill than in the sixteenth century. It is possible that some of the gente of the plantation were detailed to get wood in the eighteenth century, thus decreasing the amounts purchased, but I think that this is not likely; the explanation more probably depends on increased use of bagasse or improvements in the design and construction of the furnaces and the extraction rate of the mills.

It is not possible to fix the time at which bagasse became a principal fuel at the mill. The Governor of Morelos stated in his annual report of 1881 that all the mills used bagasse for fuel, and Haven (1875, 244) described briefly the use and handling of bagasse at Atlacomulco: "The corn-husks [sic] are dragged into the court, spread, dried, and used for fuel the next day. The fuel ready for to-morrow's burning was twenty feet high and wide, and two hundred feet long, the refuse of a single day." The nineteenth century, with its improved milling equipment, must have seen a considerable reduction in the proportion of wood used for fuel in the ingenios. The introduction of the mancuerna, a pair of kettles hung to the same fire and the nearest approach in Morelos to the West Indian Jamaica train, took place sometime early in the nineteenth century and probably also resulted in a saving of fuel.

Controversy centering on rights to cut wood for mills from the nearby woodlands began almost as soon as Tlaltenango was built, and even in the sixteenth century claims were made that wood was expensive and likely to be scarce at some future time. In the first lawsuit (L282 E4) dealing with the wood supply, the Marqués was forbidden by Royal Order to restrict cutting near Cuernavaca, and the Marqués' lawyers responded that not only was his power to license cutting in conformity with well-established custom, but that it was further justified by the fact that cutting practices had been wasteful, in opposition to royal laws. It was their opinion that the continuous and large ingenio fires would shortly strip the surrounding fifteen leagues of all trees, and in 1926 Ruiz de Velasco (p. 133) wrote that the flow of the famous line of springs of Morelos was diminished in the late nineteenth century by extensive cutting of forests for the mills and railroad.

IRON, STEEL, AND COPPER

As was the case with so many other supplies, iron was purchased lavishly in the late sixteenth century, averaging 3,050 pounds annually in the years for which data are available. In later but shorter runs of years, less was bought at lower prices. For example, the amount used annually in the 5 years 1620–24 was 894 pounds; for 4 years between 1768 and 1781, it rose to 1,586 pounds, and for 4 years between 1811 and 1822 it was only 695 pounds. The average unit cost rose in the late eighteenth century, as did the unit price of steel, but very little of the latter was used at the mill. As iron parts replaced wooden ones in the nineteenth century the payments for repairs and parts to itinerant smiths increased, but still the total average annual outlay for iron seems to have decreased greatly by the early nineteenth century from the high levels of the sixteenth century. This conclusion is evident from even casual inspection of the accounts of the two periods, which differ greatly in respect of the frequency of mention of iron, leading to the further conclusion that the iron available in the eighteenth and nineteenth centuries was more durable than it had been earlier.

Copper was not widely used at the plantation: near-ly all of it was in vessels used in the boiling-house or on the sleeves of the rollers, but at any time there was probably more copper than iron by weight at the plantation. The boiling kettles constituted the greatest single concentration of copper at the mill. The total weights of these copper kettles are not available before the inventory of 1721, when they totaled about 5,000 pounds, but probably before that date their total weight was less. By the late eighteenth century, the total weights of the kettles had more than doubled, and finally in 1884 they amounted to almost 9 tons. By contrast, for the last quarter of the eighteenth century, the total amount of copper in use and on hand at the mill, including the copper sleeves of the rollers, was about 10 tons. The total weight of copper and bronze — the latter was much less important — on hand and bought for repairs in the late eighteenth and part of the nineteenth century is available for only four years. In each case it was over a ton, and in 1769 when extensive replacement of teaches and kettles was necessary, the amount bought was well in excess of 4 tons.

Only very little information is available on the amount of copper bought annually, but it implies that the repair and replacement rates of the teaches and kettles must have been rather high. Some of the letters written by the mayordomo in the short period 1703–6 reinforce this impression (L30 V56 E1). Boiling ceased when several kettles gave out at once, as in 1769 (L341 E5), when most of the kettle bottoms had to be replaced; the repair work lasted for more than a month, from the middle of August to the end of September. In 1824 (L420 E14) it was similarly necessary to stop work, buy a large amount of copper to repair the kettles, and then to reseat them. Much of the work and expense consisted in removing the kettles and replacing them — for example, when taking inventory, it might cost as much as $500 for removal and replacement merely to weigh the kettles and note their condition.

For the sixteenth century, it is not possible to separate the amounts spent on copper and iron, since they were generally listed together in the annual accounts; nor is it possible to state the total amount spent for both, since in many years the Governor sent large amounts of both kinds of metal to the plantation and the cost was not included in the plantation accounts because no cash outlay had been made by the mayordomo. Data for only three years (1590, 1591, 1596) show that the amount charged to the plantation for both kinds of metal was approximately $1,000.

VIII

Labor

THE most satisfactory analysis of labor inputs at Atlacomulco would be based on types of work done, which may generally be described for sugar plantations in the following terms: plowing, planting, weeding, irrigating, cutting, milling, boiling, purging, and transporting. However, until sometime in the mid-eighteenth century, it was customary in Mexican mills to describe labor inputs in terms based on a very different view of labor, and this practice has left data whose nature must strongly modify the ideal analysis.

In the sixteenth and early seventeenth centuries at Tlaltenango, labor was classified primarily on the basis of race and the differing ways in which members of different races were recruited, without reference to the kind of work done by most of the Indians hired. Most Indians were either field hands, recruited by persuasion or in terms of obligations imposed on Indian pueblos in the manner of the repartimiento, or else they were classified as naboríos, persons living at the mill itself. Some of the Negro males and all of the Spaniards and naboríos were identified by the work they performed, but in the sixteenth- and seventeenth-century accounts other Indians were not so identified. Instead, only the total annual wages paid to them were recorded. Because of this practice, it is impossible to assign labor inputs to the distinct operations performed at the plantation in the sixteenth and seventeenth centuries; instead, the analysis for the period 1581–1625 must center on rates of pay, length of association with the plantation, and average daily output of sugar per man and day.

By contrast, inputs according to type of work done may be calculated from data in the eighteenth-century accounts, because of marked differences in their form (see Appendix F, p. 116). Thus, daily and annual outputs and costs of labor inputs afford the only points of comparison with the data of the period 1581–1625, and it is impossible to see which kinds of operations experienced the greatest gains in efficiency.

Spanish Labor at the Plantation

Upon scanning the plantation records, one's first impression might be that many Spaniards worked for long periods at the plantation, making it appear that continuity was an important characteristic of this part of the labor force. In fact, this impression is erroneous, and may be due largely to the fact that a few persons were indeed associated with the plantation for long periods; on scanning the records their oft-repeated names fix themselves in the memory and create an illusion of continuity. Instead, large numbers of Spaniards occupied the skilled and supervisory positions at the plantation, and the rate of turnover might more correctly be described as dizzying.

I have tabulated below the length of service of Spaniards occupying all posts except that of mayordomo at Tlaltenango for the years from 1581 to 1625 to provide a firmer base for discussion, dividing their occupations into two groups: those having an important element of record-keeping as a characteristic, and those which did not. In the first group are the five positions of sugarmaster (maestro de azúcar), purger (purgador), dispenser (dispensor), priest, and doctor. In the second group are the eight positions of oxherd (boyero), muleteer (harriero), carter (carretero), cane supervisor (cañaverero), cane guard (guardacaña), muleherd (guardamulado), recruiter of Indians (recojedor de Indios), and overseer (mandador). Data are unavailable for only 3 of the 44 years from 1581 to 1624 concerning the Spaniards who worked at the mill.

	Records	Non-Records	Total
Number of persons	62	86	148
Mean tenure	2.6	2.9	
Worked >5 years	5	12	17
Worked 1 year or less ...	33	49	82

The mean length of tenure in the first group of occupations was only slightly less than that for the second; relatively fewer in the first worked more than five years at Tlaltenango. By contrast, more of the second group stayed a short time (one year or less) than did members of the first. More than half in both groups worked only one year or less, another third worked slightly less time than the average for each group, and in fact the 17 persons who worked for more than five years—in many cases intermittently—raised the average length of service to figures unrepresentative of the experience of most workers. Roughly speaking, the probability that anyone hired would work for a year or less was 0.5, and at the extreme 18 persons stayed a month or less. The Spanish labor supply was unstable, but at the same time it was not difficult to fill positions in this period when dependence on Spaniards was uniquely high.

Mayordomos. At the apex of the plantation hierarchy was the mayordomo, directly responsible to the Governor. He was guided in part by instructions from the latter; although little of the correspondence between them has survived, the amount available suggests that its frequency was high, each writing as many as two letters per week or even more. The mayordomo apprised the Governor of current conditions and needs for supplies and money, and the Governor in turn sent instructions and at least some of the items requested. The mayordomo received a salary, an allowance for food nearly as great, living quarters, and, in most years, an incentive payment based on the amount of profits of the plantation, usually 5 per cent.

His workload, characterized by the Governor in 1808 as "recio" at a time when it was difficult to fill the post (L219 E1) was more fully described by Lucas Alamán in 1825 (L219 E3) when he was attempting to persuade the Junta not to lower his salary. Alamán wrote that from 5 to 8 A.M. the mayordomo planned and assigned the work of the day, checked on fieldwork from 8 A.M. until noon, then worked in the office until 3 P.M.; after reviewing the fieldwork from 3 until 6 P.M., he worked on the daily labor and production accounts until 9 or 10 P.M. This daily schedule of 16 or 17 hours occupied at least six days of the week. Although labor inputs at the plantation experienced marked seasonal variation, it seems unlikely that the mayordomo enjoyed a lighter load in the slack times, since the range of activities he supervised, as well as the distance he covered in moving about the plantation, was so great. Nevertheless, his salary and allowances amounted to a fairly good income in most years.

Histories of the Mayordomos. Appendix G (p. 118) contains all the important details that I have found concerning the mayordomos. In contrast with the Governors and the renters of the plantation, most of whom became involved in lawsuits with the Estate for various reasons, the mayordomos seem to have been much less troublesome. The major source of their difficulties was not direct misappropriation of funds, but rather accounting for the materials and livestock charged to them at the beginning of their tenures. Discrepancies required explanation, even in cases where livestock were involved and some reasonable amount of annual loss might be accepted. I have found very few cases where the amount of loss charged against the mayordomos was excessive or did not have a reasonable explanation.

Dispensers and Purgers. I have grouped these two positions together because often the same person occupied both posts in the same year, or was described as dispensero in one year and purgador in the following, and it appears that the two positions were not sharply distinguished. The incumbent issued supplies and kept records of sugar and molasses production. The importance of this post, second to that of the mayordomo until the nineteenth century, was such that I have included information about some of its occupants in Appendix G (p. 118).

Priests. A priest was in residence from the beginning of the annual accounts in 1581 until at least 1612; from then until 1616 there is no mention of a priest in the salary lists, and beginning with the account of 1616–17 there appear notices of payments given to the Convent of San Francisco of Cuernavaca for "doctrina de la gente." A priest was never again in residence.

Although there is little information concerning the priests, some appeared to have occupied it for relatively long times — 4 or 5 years — in contrast to a few who remained only a very short time. I have identified only two who had business ties with the plantation in addition to performing religious duties there. Gaspar de Praves, bachiller, grew cane that was processed at Tlaltenango on shares in 1598; cane grown by him continued to be so favored until at least 1608, so that his association with Tlaltenango extended beyond his residence there. Marcos Millán de Velasco (1604–7) sold nearly 1,000 loads of wood to Tlaltenango in 1606, and in 1607 was paid $300 for some seed cane raised by him on land belonging to a resident of Jiutepec. These two are the only documented examples of persons with local economic interests, but three others, Juan de Ayllón, Hernando de Castro, and Bartólome de Cabrera, shared family names with other persons who either worked for the mill at approximately the same times, or who sold wood to it.

In addition to priestly duties, the plantation priests certified births and deaths, as well as the causes of the latter; in 1602–3 the vicario kept records of sugar made at the mill, but after priests ceased to be in residence they performed only religious duties, including the recording of births and deaths.

Médicos, Barberos, and Cirujanos. Of the persons who supplied medical attention in the late sixteenth and early seventeenth centuries, only three men seem to have stayed a relatively long time, and of these only one, Juan de Mendoza, filled other positions for the Estate and carried on local business activities. From 1600 to 1625 the plantation was not without the services of a doctor any great length of time, but after that consultation rather than residence became customary. It might be wondered whether extensive training was regarded as necessary for some medical activities; for example, Damian Marqués was a vaquero at Tlaltenango before he was called upon intermittently to let blood.

Overseers. Of the 29 persons identified as overseer (mandador) of field operations, well over half (18) remained in the position for a short time, appearing in annual accounts only once or twice. Of the others, associated for longer periods and with a wider range of employment — selling wood or cane to the mill, transporting sugar, occupying other posts at the plantation — perhaps the most interesting is Gregorio de Figueroa, whose association with Tlaltenango continued from at least 1581 to 1613, more than thirty years. A major feature of the position of mandador was its frequent combination with the position of recogedor de Indios —

that is, the mandador might find Indian labor as well as supervise it. Figueroa must have developed important local ties that enabled him to accomplish the task, inasmuch as the large amounts of wood that he supplied to the mill over a long period were probably bought from Indians. His career also illustrates another feature of overseers who served a long time that further distinguishes them from other Spanish workers: they occupied a wider range of positions than did others, functioning as oxherders, muleteers, canefield supervisors, labradores, and, in Figueroa's case, as carretero and smith as well as overseer.

Cañaverero, Labrador, and Guardacaña. The first two positions were sharply differentiated by salary until at least 1600; in that year, for example, Gaspar Martín received an increase in annual pay of slightly more than $100 when he became cañaverero after working more than a year as labrador (L263 E15). In 1603 he was identified as cañaverero y labrador, and continued in this dual position until 1608, when he died shortly after becoming dispenser (L268 E3). The position of labrador was fully described in 1603, when its holder was charged with preparing land for planting and covering the seed cane after it had been placed in the furrows; earlier, in 1599, the phrase "labrador de las sementeras" had been used. The work of cañaverero began after this step, since he was responsible for guarding the cane — fires and livestock offered the principal hazards — as well as supervising weeding and irrigation. Ten fewer names are listed for the post of labrador than for that of canefield supervisor, and the periods of service of labradores were short; one exception worked 31½ months, from 1587 to 1589. The latest date at which the occupation was identified is 1608, when Gaspar Martín was both labrador and canefield supervisor. The latest date when anyone occupied only the post of labrador is 1602, so the distinction ceased to be important in the first decade of the seventeenth century.

The importance of the position of cañaverero is shown by the fact that it was continuously staffed during the period 1581–1625; in fact, in some years as many as three persons were canefield supervisors simultaneously, although only for a short time. Although at least 24 persons were identified as cañaverero, four persons occupied the post most of the time between 1581 and 1625, and all except one worked at other tasks on the plantation as well. Of the four principal tenures, three were ended by either death or rental of the plantation.

Information concerning the guardacañas is available

for only the decade 1599–1609. Six names occur, of which two are found in other accounts as minor sellers of wood to the plantation, and the rest occupied the office for short times — probably in no case did any guardacaña work for more than 4 months — and received only one mention each. This post ceased to offer opportunity to Spanish labor at about the same time as did the post of labrador: the latter was not mentioned after 1608, and the former disappeared after 1609.

Muleteers. Only eight persons are identified as muleteers (harrieros) in the accounts, and all except one began their employment before 1589; the single exception worked intermittently from 1615 to 1625. He was not salaried as had been previous harrieros, but rather was paid by the task (L269 E1). Thus the salaried position of harriero was abolished about 1590, and from that date transport was supplied by local persons, many of them plantation employees, who used their own mules to carry goods and supplies. Of the four who were long-term employees of Tlaltenango, one died after 5 years of intermittent work, two appear in the accounts over a period of 10 years, and the fourth for 20 years. All of them occupied at least one other position at some time at the plantation.

Carters and Oxdrivers. Of these two related positions, the most important appears to have been that of oxdriver or oxherd (boyero), since it was occupied continuously in the period 1581–1625, whereas the last mention of a Spanish carretero occurred in 1602, when it was occupied for the year by the long-term worker Alonso de Santayana. Only four persons were designated carreteros, and three of them worked for long periods, in contrast with the oxherds, most of whom worked only for short periods.

Sugarmasters. The importance of this position is made obvious by the high salary attached to it. It was approximately equal to the mayordomo's salary and more than double that received by Spaniards in other positions. Information from the years 1581–1625 shows that for the nine persons who held the post it was distinctly not a part-time occupation; even those who held the post for only one or two years received the full annual salary. Of the nine occupants, two worked for long periods: Juan Fernández de Mata (1582–90) and Antonio Rodríguez de Cárdenas (1593–1603). Fernández grew cane on his own that was milled at Tlaltenango, sold wood to the mill, and carried cane to the mill and sugar to Mexico City on his own mules (L257 E13). By contrast, there is no indication that his principal successor, Rodríguez, was engaged in this common range of activities, nor was the highly paid Simon García, who was earning $1,000 annually as sugarmaster when he died. Of interest is the adjustment of sugarmaster salaries to the fall in sugar prices at the turn of the century, and the fact that two of the last three sugarmasters occupied other posts at the plantation: Juan de Rojas alternated supervision of sugarmaking with the practice of medicine, and Diego Felipe was alternately overseer and sugarmaster.

The last sugarmaster concerning whom there is information from the period 1581–1625 was a free mulatto, Diego Martín (L269 E18), whose racial background is of interest because through the rest of the seventeenth century and much of the eighteenth Spaniards were not employed as sugarmasters while the mill was administered by the Estate. In fact, much of the time slaves occupied the position, and several eighteenth-century inventories identify one or more slaves as sugarmasters. The substitution of Negroes and mulattoes for Spaniards in this position occurred even though Spaniards continued to be hired as purgadores, possibly because the latter work required at least semi-literacy.

Smiths, Carpenters, and Potters. Apparently the office of blacksmith (herrero) offered more opportunity for full-time work in the period 1581–1625 than did the position of carpenter (carpintero); three smiths of the total of nine worked full time for more than a year, whereas only one of seven carpenters worked nearly full time — his tenure, however, extended over the period 1588–91, one year less than the longest term occupied by a smith. It is clear that most carpenters, like other specialists such as kettle repairmen, were hired for specific jobs even before 1625; after this date all such work was intermittent. Before 1625, two of the carpenters fixed only carts, and two others repaired cane-crushing equipment only. Of the 16 individuals within these two occupational categories, only two worked for the mill in other positions: one, the mulatto Baltasar García, became overseer in 1615 long after his intermittent work as smith from 1581 to 1589, while the other worked briefly as carretero. Only one of the sixteen sold wood to the mill (L262 E1); the lack of varied economic activity by individuals, including Spaniards, in this fairly large group was probably related to the rather low salaries they received. In this connection, it is of interest to note that the lone Spanish potter, hired to make the clay forms used in purging, was paid much more than either smiths or carpenters in 1581: his annual salary was $300, compared with $230 for a smith and $150 for a carpenter.

COSTS OF SPANISH LABOR AT TLALTENANGO

Table 27 (p. 133) shows the major features of the history of salary payments to Spaniards working at the plantation. Only the payments made to the mandadores form a reasonably complete series, but in this case complications are introduced by the practice of paying different persons at different rates in the same year for the same work. However, the table does show that from 1581 to 1593 a common annual salary for a mandador was $200, and that the payment increased first in the 1590's and again in the first decade of the seventeenth century. Perhaps the first increase occurred more or less simultaneously with an increase in the annual ration allowance from $24, standard for most Spaniards, to $41; it also appears that the salary increase for mandadores that occurred in the first decade of the seventeenth century was accompanied by yet another increase in the ration, from $41 to $48. From this peak in the early seventeenth century, salaries dropped in the second and third decades, returning for mandadores to the levels existing about 1590. The series of mayordomo salaries shows that further declines occurred until about 1700, after which the annual payment remained at $500 until the War of Independence in the early nineteenth century; the salary for mandador, by contrast, had apparently achieved relative stability by mid-seventeenth century, remaining at about $150 until sometime after 1831, or for approximately two centuries.

The table also shows clearly the marked and important reduction in dependence on Spanish personnel that occurred in the early seventeenth century. From this time on, probably only the administrator and the purger were Spanish, whereas other administrative and supervisory personnel may have been mestizo, mulatto and Negro slaves and freemen, a fact of extreme importance in reducing salary costs, a major item of expense in the sixteenth century.

Negro Labor at the Plantation

Spaniards were fond of remarking that Negro slaves formed the core of the labor force, and it is true that they dominated in mill and boiling-house, in accordance with law; contrasted with the total inputs of Indian labor, however, the importance of Negro slaves at the Cortés plantation occupied second place.

The first large importation of Negro slaves by Cortés for use on his plantations occurred in 1544, when a shipment arrived for the Tuxtla mill. Cortés had contracted with Leonardo Lomelín, a Genoese, on 11 May 1542 for the shipment of 500 Negroes from the Cape Verde Islands, two-thirds to be male and the rest female, between the ages of 15 and 26 (L270 E4). A receipt for one shipment of 100 in fulfillment of the contract shows that Cortés' agent refused 2, and described most of the 98 acceptable Negroes as thin and tired (L247 E8). Delivery of the next large shipment concerning which there is information occurred in January 1579, when 40 were sent via Spain — only 5 were females — to the mill at Tuxtla (L247 E10). In later years the lots were much smaller; one major lot in 1598 contained only 19 individuals (3 females), and the major feature of later purchases of imported slaves remained their small size. Many were bought in Mexico City and had spent time there before being sent to the plantation. Information in Table 24 (p. 133) demonstrates the decrease in reliance on imported slaves between 1579 and 1623.

After 1623, purchases of small groups or of individuals were common: in 1655 and 1656, 21 slaves were bought from as many different persons in Mexico; 19 were mulattoes born in Mexico, and only 2 were identified as African. Subsequent purchases in the 1670's and 1680's (L93 E2) were of small groups containing hardly any persons born in Africa, and many were mulattoes. Whether this changing pattern of purchase at Atlacomulco duplicates a more general pattern for Mexico I do not know; Aguirre Beltrán, who has provided the major source of data on the Mexican slave trade, was unable to provide information by years or longer periods of slave imports in the late seventeenth and eighteenth centuries that might be compared with experience at Atlacomulco.

The Atlacomulco data are nevertheless adequate to discuss the long-term relative importance of purchase of slaves and their acquisition by birth on the plantation. About half (445) of all the slaves concerning whom I have information were born on the plantation, and their numbers were approximately equally male (218) and female (227). Over the long run of the slaveholding history, therefore, it might be useful to state as an approximation that for every slave bought, one was born at the plantation. Since many of the latter died before reaching maturity, however, it is obvious that purchase of slaves was more important than local reproduction in maintaining the work force.

Place of Birth of Slaves. It was customary to include in inventories information concerning whether the slave had been born in Africa or was a creole (criollo); thus it is possible for the 889 persons dealt with in Table 25 (p. 133) to classify all but 35 by place of birth. Most (33) of the 35 not classified in this way were males who had spent at least some time in Mexico in the ownership of some other person before purchase by the Estate, and nearly all of these had Spanish Christian names and surnames.

Of the 889 individuals in the group, approximately one-third were Negroes born in Africa, and it appears likely as well that more than half of all 889 were full-blooded Negroes, the difference representing creole Negroes. Table 26 (p. 133) is a summary of data concerning the places of birth of slaves from Africa, and is primarily confirmation of Aguirre Beltrán's generalizations about African slave sources; thus the picture is conventional for sixteenth- and seventeenth-century New Spain. Half came from the Senegambian coast and interior, where the Biafara, Zape, and Mandingo groups were major sources. Groups yielding three or fewer slaves were the Biojo, Xoxo, Cabo Verde, Cazanga, and Balanta. Approximately half as important numerically were slaves brought at a later date from the Congo basin and coast and from Angola. Most of the latter were identified simply as from tierra Angola; tribes providing only one or two slaves were the Mocanga, Malemba, Matamba, and Xigo. Somewhat less important was the region to the north, comprising the coastal region between what are now called Ghana and Río Muni, and including São Tomé. Only 7 were born along the east coast of Africa. Of the 27 slaves classified as "Other," most were brought to Tlaltenango from Tuxtla in 1587 and no record was kept of their place of birth; in addition, this group includes some from the entrepôts of Santo Domingo and Portugal, as well as a few whose place of origin was not listed in the work of Aguirre Beltrán and which I cannot identify.

Importance of Racial Mixture. More difficult than identifying place of birth is the identification of an individual's racial heritage. This is largely due to the fact that the sixteenth-century inventories describe the offspring of African-born slaves as criollo, perhaps because most of the first and second generations of slaves born on the plantation were in fact full-blooded Negroes. Not until the inventory of 1585 were mulatto children identified, but this cannot be taken as proof that none existed before the 1580's.

Table 25 shows that persons who were identified as mulattoes or who had at least one mulatto parent numbered only half the total number of Negroes, yet there remains the large group of 229 criollos that may have contained sufficient mulattoes to raise their numbers to one-third of the entire slave population.

The proportion of Indian-Negro crosses within the mulatto group, in contrast to Negro-Spanish or Spanish-Indian, is difficult to fix because of the wide range of meaning of the word "mulato." According to Aguirre Beltrán (162), the Indian-Negro mixture predominated over the others, and although the word "zambaigo" designated this type it was little used, being rapidly replaced by "mulato," which was also applied to persons of Indian and Spanish blood mixed with Negro. I have seen the word "zamba" used in reference to only one of these approximately 900 slaves, and I assume that it refers to mixed Indian-Negro ancestry. Instead, adjectives such as cocho, prieto, and alobado were combined with "mulato" to describe Indian-Negro mixtures, and the term "mulato blanco" applied to persons of Negro-Spanish descent (*ibid.*, pp. 167–69). The blanco group had only 11 identified representatives at the plantation, a very low figure, and four times as many persons had Indian and Negro ancestry. Much more numerous than either of these two groups were persons described merely as "mulato," and it is possible, following Aguirre Beltrán, to assume that most of these were predominantly Negro and Indian.

Since the purchase of mulattoes was uncommon, the members of this group were almost entirely native to the plantation. In fact, I have been able to determine that only 9 mulatto women and 24 men were purchased, most of them bought in the mid-seventeenth century by Governor Valles (2 women and 17 males) and in 1721 from the small hacienda Atotonilco (6 females and 4 males). The "mulato" group inevitably became dominant as time passed, since the word was applied to so wide a range of racial mixture, and slaves were no longer purchased on the scale that they had been in the sixteenth and very early seventeenth century.

Total Numbers of Slaves. An overview of the slave-holding history of the plantation is given in Figure 17B, which shows that Negro slave labor was important numerically only in the second half of the sixteenth century and the first two decades of the seventeenth. Peak numbers were held in 1566, when there were 155 slaves described in the inventory, two-thirds of them male. An effort was made to obtain slaves to comply with the late

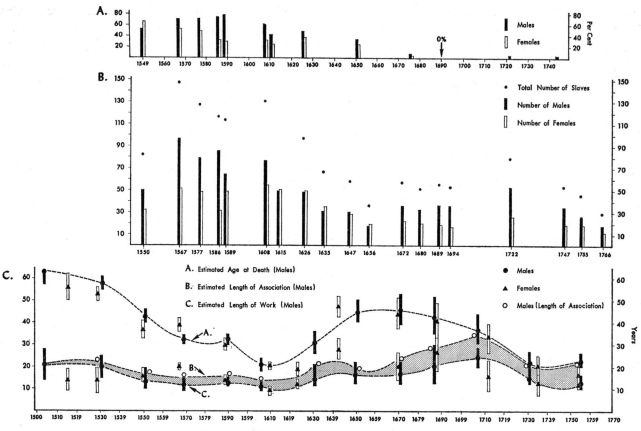

Figure 17. Vital statistics of the slave population

sixteenth-century edict prohibiting the use of Indians in millwork, allowing their use only in the fields, but important purchases had already been made by 1596 before the edict, probably as a result of decrease in the Indian labor supply. After 1607 the slave population declined, reaching a low in 1655, but a few purchases and reproduction kept the population between 50 and 60 until about 1700. A major purchase was made in 1721, with the acquisition of all the slaves of Atotonilco. No attempt was made after this to acquire many other slaves nor even to recapture the many who fled, an important source of loss in the middle eighteenth century, with the result that finally the small slave population consisted primarily of the very old, the young, and the sick and disabled, and probably made little contribution to the work of the plantation. Nevertheless, because many of the few slaves in the late seventeenth and eighteenth centuries were specialists in various branches of plantation work, including such key posts as sugarmaster, and could do the work of salaried Spaniards, their contribution to the labor force was important.

Birth and Death Rates of Slaves. The custom of tak-

ing frequent inventory of the plantation produced many lists of names of individuals, including in many cases their ages, information concerning relationships, and descriptions of work specialties. In addition, births and deaths as well as purchases and sales of individuals were itemized in numerous annual accounts. This information, combined with the few remaining examples of lists accounting for the distribution of weekly rations and the more numerous lists accounting for distribution of the annual clothing allowance in some accounts, enables fixing with varying degrees of precision the birth and death dates of 740 (slightly more than 80 per cent) of the total of 916 slaves. The latter figure is a minimum but nearly complete figure of the total number of slaves held between 1549, the year of the first inventory, and about 1800. The information also permits estimation of the years of work contributed by 833 slaves (nearly 90 per cent of the total), and approximately the same number are included in the estimates of length of association with the plantation, which include childhood and old age as well as working years; these categories are discussed separately below.

The most important problem offered by the data concerns the fates of slaves introduced in the Lomelín shipments: as can be seen in Figure 17C, the average age of death of slaves alive at the time of the first three inventories was extremely high, yet there is no information concerning those who died between 1544 and 1549 — that is, the mean figures refer only to slaves who had survived the first few (up to five) years of residence in the New World. Furthermore, the addition from Tuxtla of survivors of these earliest shipments also operated to overstate the average age at death of those born before 1550, since there is no information on mortality rates at that plantation.

From the 1549 inventory, I have constructed the following tabulation to show the distribution at Tlaltenango of age groups of Negro slaves in 1549:

Age	Male	Female
0–6	7	11
7–15	0	0
16–25	3	2
26–35	9	8
46–55	9	0
56–65	0	1
66–105	0	0
106–110	1	0

The 18 persons aged 0–6 years were offspring born in the New World, and there were none in the age range 7–15. Probably all the rest were survivors of the Lomelín shipments, and in fact the ratio of males to females (34:19) over 21 years of age is approximately 2:1 as specified in the contract. However, the stated ages of most of the persons listed in the 1549 inventory are much too high to have fulfilled the condition of the contract that stipulated ages of 15–25 on delivery: 34 of the persons over 21 were older than 35 in 1549, so it is possible that there was a tendency to exaggerate ages owing either to rapid aging under plantation conditions or to some other consideration associated with the evaluation process; on the other hand, some slaves may have been introduced from other sources.

Information concerning the survival rates of imported slaves is limited to the histories of four shipments: one of 40 slaves to Tuxtla in 1579 (L247 E10), and three to Tlaltenango containing 47 Negroes in 1598, 1599, and 1600. For 8 persons, there is no information; of the rest 8 died within 7 years, 60 after 7 years, and 1 fled the plantation. The percentages of slaves lost within the first 7 years for each shipment were 18, 5, 30, and 40, and the average of all four shipments was 25 per cent, if one assumes that the 8 persons concerning

whom there is no information died within a short time of arrival. If the same rate of loss is applied to the period 1544–49, then it appears that fewer than 100 slaves were received from Lomelín for Tlaltenango, which may well have been the case, inasmuch as the plantation was able to draw on a still large reservoir of Indian labor. It is also of interest to note that in the inventory of 1607 the ages assigned to the slaves received in the three years 1598, 1599, and 1600 were from 5 to 10 years greater than they should have been had the ages assigned on arrival been correct, a tendency consistent with that displayed in the 1549 inventory.

The reality of the problem of ages of imported slaves is illustrated by the data contained in L247 E7, wherein are listed the causes of death between 1572 and 1586 at Tuxtla of many slaves, other than those recently imported from Africa. By 1584, some of these slaves might have been 65 had the clause specifying ages in the Lomelín contract been fulfilled, but of the 31 individuals listed, 13 were described as 70 or older, including a woman over 100 and a man over 90. It is not easy to dispose of the problem. The mayordomo was necessarily concerned to overstate ages because he needed eventually to demonstrate that death was due to natural causes rather than to mismanagement; in the case of inventories, usually taken by at least 4 experienced persons of the vicinity appointed by each of the parties, it is not clear why overstatement should have been the rule, apart from physical appearance suggesting rather rapid aging. Nor did the slaves know their own ages. However, I have found that in cases where it was possible for the assessors to fix the date of birth or baptism of slaves born on the plantation, the tendency to overstate ages was less extreme.

Figure 17C shows long-term changes in the average age at death. The most conspicuous features are the great drop occurring over the first century of the period under review, to a low point for slaves born in the decades 1600–19, followed by a rise to a secondary peak of life expectancy for persons born in the decades 1650–89; the subsequent decline brought the life expectancies of those born in the decades 1720–39 to the low level of those born in the decades 1600–19, after which the very small number of slaves at the plantation does not make generalization worth much.

As stated above, the life expectancies of those born in the decades 1500–39 must represent overestimates because they are based on the statistics of a population of survivors in 1549. Nevertheless, a decline in life expectancy probably did occur, for two important reasons: the slaves brought from Africa were selected for at least the appearance of health, and apparently successfully selected on this basis; furthermore, the number of children born to the slave population seems to have increased after 1560, and since the infant mortality rate was high, the inclusion of increased numbers of births decreased average life expectancy. Given these circumstances, it is interesting to see that after reaching low values for those born in the first two decades of the seventeenth century, life expectancy increased significantly at the same time that the percentage of all slaves born in Africa, hence selected for potential longevity, decreased markedly (Fig. 17A). That is, the increase occurred in spite of the fact that few persons were added to the population who had survived the hazards of childhood elsewhere than New Spain, and the small purchases of locally born persons included few full-blooded Negroes.

The causes of changes in life expectancy must be approached indirectly in the absence of information concerning causes of death. I have chosen to compare slave and Indian mortality rates in an effort to see if the well-documented epidemics that affected the Indian population in the sixteenth and early seventeenth centuries also caused changes in the slave population. To this end, I have placed on Figure 2 (p. 10), showing Indian population changes in Morelos, the total number of Negro slaves at the plantation at various times between 1606 and 1765. Between 1606 and 1650 the decline in slave numbers corresponds almost perfectly with the decline of tributaries in the nearby town of Yautepec, but since the slave population received individuals by purchase and lost them through sale, whereas Yautepec did not, the graphic comparison cannot be used alone. In addition, it is necessary to compare increases in the slave population through purchase with losses from various causes; this may be done only for the period 1609–25. Forty persons were added through purchase, 27 lost by sale, transfer, flight, and granting of freedom, leaving a net increase of 13 persons. In the same period, 56 were born on the plantation, and 113 died, leaving a net decrease of 57. Thus, the addition through purchase of approximately one person annually was insufficient to overcome net losses by death in the same period. Without additions by purchase, however, the rate of decline of the slave population might have been greater than that of the regional population, but interpretation is complicated by the

imbalance in numbers between males and females that persisted from the mid-sixteenth century.

Assuming that the epidemics that affected the Indians affected the slaves as well, I have analyzed the data by months for the period 1 January 1610 to 31 December 1623 (except the year 1612, in which year there were either no deaths or the data are missing) to see if deaths were concentrated in time. Only 10 deaths were not recorded by date in this period, whereas 113 were recorded by date. The greatest mortality occurred in the first year, 1610, when 26 persons died, 13 of them in January and February. It appears that some contagious disease was involved, since if we subtract this number (13) from the annual total (26) the deaths occurring in 1610 approach the range of the rest of the years, which was between 4 and 11. No month emerged clearly as representing a peak in mortality, if we subtract 5 deaths each from the epidemic months of January and February 1610, bringing these two months within the range (5 to 12) of all other months. Months having fewest deaths in the 14 years were August and September, with a total for the entire period of 6 and 5; otherwise, no seasonal pattern is discernible, and this one is only weakly developed. I have not tabulated the stated causes of death, since too few deaths were explained.

It might be expected that the inventory and sale prices of slaves would follow changes in life expectancy, or in expected years of work, but in fact a preliminary graphic analysis using Figure 17C as a base showed too little correlation to warrant further analysis; there exists the possibility that the changes in life expectancy were not recognized by the buyers. Instead, the average value of all slaves fell after 1607 to remain at a plateau between 1613 and 1693, after which began a persistent decline in value that characterized the eighteenth century until at least 1764. It appears, rather, that the price of slaves followed the price of sugar, with the rate of decline in the values of both being especially noticeable between 1746 and 1764.

Average Years of Work. Calculation of the average number of years at work at the plantation was based on the assumption that most persons were counted in the work force between the ages of 10 and 60 years. There are numerous items in the inventories that show 10-year-old boys, for example, to have been carreteros or irrigation assistants; in addition, the 1575 rental contract (L432) states that all adult slaves were valued at $400, and all under 12 were valued at $300, suggesting that a more or less important contribution to work was made at the age of 12. At the other extreme, it seems a reasonable assumption that some work was performed until the age of 60, although Figure 18 indicates that values began to decline at approximately 50 years of age. In all cases where data concerning the working ability of an individual were available, the figures for the individual years were adjusted in the light of such information. Although, as with the figures showing average age at death and the length of association with the plantation, it is not possible to state precisely the lengths of time involved (the calculations resulted in a range of time expressed in the form, for example, of 36 ± 5 years), I have not indicated the ranges on the graph for length of association, since it is nearly identical with the range of average age at death (line A in

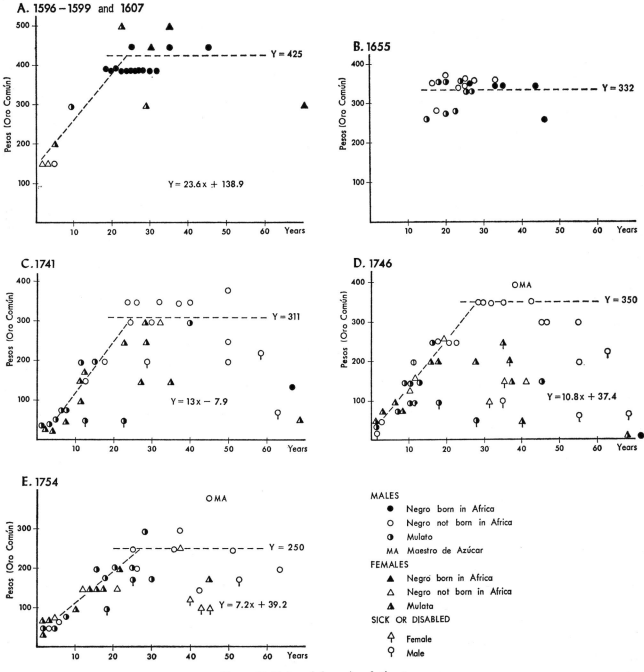

Figure 18. Value of slaves in relation to age

83

Figure 17C). In nearly all cases, the average number of years at work has been estimated within a much smaller range, as Figure 17C shows.

For persons born in the first four decades of the sixteenth century, the average years at work were about 20. The average working life then declined, finally stabilizing between about 10 and 15 years for persons born between 1550 and 1620. Over the long period from about 1620 to 1710, average working life increased, reaching a peak of slightly over 20 years for those born near 1710. Following this peak, average working life declined again.

I have included data concerning average length of association with the plantation (line B in Figure 17C) because it has a bearing on questions concerning the profitability of slavery. The difference between lines B and C is the length of time that the average individual was supported by the plantation but not working. The difference was rather small until, beginning with individuals born about 1670, it became almost half as great as the length of the working life. The resultant increase in the costs of support at both ends of the working life thus worked against the gains of a longer working life that characterized the late seventeenth century.

Discipline of Slave Labor. The most interesting information I have found concerning discipline is contained in a letter of Lucas Alamán (*Documentos,* Vol. 4, p. 533) describing the cholera epidemic of 1850 and its effects in Morelos:

En Atlacomulco no hizo mas que pasarse, pues conservandose en las haciendas de azucar el sistema monastico establecido por los espanoles, que es menester a todo trance mantener, los empleados no solo no hablan, pero ni aun levantan los ojos adelante del administrador, y bastaria que hubiese un dependiente que no pudiera sujetarse a esa severa disciplina para que relajase en todos. . . .

Thus it appears that all employees, slave and free, Negro, Indian, and mulatto, were subjected to the same rigorous discipline; even Independence did not bring release from this rule. However, the plantation records themselves do not speak concerning this point, and in fact I find it very difficult to place discipline on a scale ranging between extreme permissiveness and extreme harshness.

All the colonial inventories list items used to discipline slaves: stocks, collars, chains, leg-irons. Many inventories contain descriptions of slaves who were difficult or chronic runaways. The data give the impression of a somewhat turbulent group of workers in frequent conflict with the mayordomo, but both sides had the opportunity to appeal to the Governor of the Estate to uphold their cause. Although to the slaves the Governor must have been rather a remote figure known only through occasional visits to the plantation that were accompanied by a certain amount of ceremony, they could and did appeal to him when they felt they were mistreated; the mayordomo, on the other hand, could invoke the distant Governor as arbiter in cases that he felt required severe discipline without himself appearing to have full responsibility for the final decision (L240 E4, 1591; L241 E3, 1594).

Some show of discipline was intended to forestall poor behavior: de Ribaguda Montoya wrote to the Governor in 1592 (L245 E1) when the milling was about to begin that he intended to gather the Negroes together the day before milling began to speak to them about proper behavior and performance. His talk would contain the information that they were to obey the Governor, that they would be treated well if they obeyed, but punished "con rigor" if they did not. To dramatize his intentions, he would select a Negro with whom he was extremely annoyed and give him four lashes: "Beamos si con esto los puede llevar." Complaints about the difficulty of controlling the slaves are so numerous in the few surviving letters of the mayordomos that one can only conclude that their relations were one of the most unpleasant features of the work as far as the mayordomos were concerned.

On the other hand, the Negroes complained to the Governors about the mayordomos, and could take legal action against them. In 1738 (L79, *Hojas Sueltas*) María Theresa, parda esclava of Atlacomulco, brought suit against Joseph Santos, a muleteer and vecino of Cuernavaca, concerning the "extrepe que cometio" against her daughter. She wanted him in jail, "lo que no haciendo efecto por resistirlo Don Manuel de Mesa, Administrador [of the mill], quien se opone a ello solo con el animo de que mis pretenciones no tengan efecto por la grande enemiga que me tiene a causa de estar siguiendo otros autos contra el susodho por la misma causa. . . ." The judge ruled in favor of María Theresa and ordered Santos to jail.

Flight offered an apparently easy solution. In this case, however, the slave's side was not likely to be heard by the Governor; in fact, I have seen no explanation apart from comments by the mayordomos intended to minimize their own responsibility for events. Bárcena wrote in a letter of 18 March 1719 (L30 V57 E3) that "abra un mes mas o menos que se me huyo sin mas causa que su gusto Joseph Antonio Mulato Blanco Mansebo." The mayordomo Urtado wrote pointedly to the Governor on 27 November 1702 (L30 V56 E1) that two slaves had fled whom "Vmd. mando desaprisionar." In other cases (*ibid.*) the motive of flight was said to be robbery. There were numerous bands of highwaymen consisting of escaped slaves, mulattoes, and Indians, particularly on the road to the capital, attracted by shipments of money used for weekly salary payments at the haciendas.

The data on frequency of flight from the plantation are inadequate, since most are found in inventories and data from annual accounts suggest that reliance on the inventories results in a marked underestimate of the real frequency. Slaves described as runaways in the inventories were in general never found, but many of the annual accounts describe the costs of recapturing slaves who managed to escape for only a short time. Thus, while the inventories describe only 7 slaves as runaways in the two decades 1580–99, in 1581 alone, according to the annual account, at least 20 slaves fled but were recaptured. In part of 1584 and 1585, 13 fled, and in 1594 it cost $191 to bring 3 runaways back from the Mixteca. According to the inventories, at least 67 persons escaped permanently or for periods longer than a year between 1544 and 1799, a figure that must be much lower than the number of slaves whose flights were of short duration.

The distribution in time of the 67 more successful flights is significant. Before 1699, 27 persons were described in inventories as runaways, but 35 were so described in the much shorter period 1740–99, 19 of them in the two decades 1740–59. These data substantiate the impression, readily gained from scanning the records, that not only were there more runaways in the mid-eighteenth century than at any other time, but less trouble was taken to capture them than formerly. Thus, of the 7 males described in inventories as captured after having fled, all are mentioned in inventories belonging to the decades 1580–1679, and I have found no reference to recaptured slaves after the latter date. Part of the explanation of the increase in successful flight may be that it was easier for the predominantly mulatto slaves to disappear into a larger free mulatto population. The decrease in the numbers of males who fled after 1759 is due to the simple fact that there were very few males left on the plantation. Many who did

not flee were physically handicapped in some way, but on the other hand some handicapped males did take flight.

Mention in inventories of slaves in irons or with scars from irons also changed. In the years from 1580 to 1659, seven males and one female had such marks, but none are so described after the latter date, and eighteenth-century inventories show a decrease in prison equipment.

General Attitudes toward Slaves. The rental contracts generally contained a clause stating that slaves should be treated well and fairly, punished neither excessively nor harshly. Although the owner was expected to bear the risk of death by natural causes, the renter was held responsible for the value of slaves who died as a result of harsh punishment or excessively hard work. The governors instructed the administrators to take good care of the slaves, and in numerous instances the administrators found themselves obliged to write to the governors for supplies — particularly clothing — for the slaves.

In their letters to the governors, the mayordomos never praised any of the workers, Negro, Indian, mulatto, or Spanish, for the quality or amount of work done. Instead, they complained frequently about the difficulty of managing the entire labor force. In their view, the slaves' attitude toward work was a racial trait, as is shown by this excerpt from a letter of recommendation written by Cristóbal Bárcena to the Governor concerning one Domingo de la Cruz (L30 V57 E1; 24 November 1719): "Un mulato cariaco y otado. . . . Es de profesion vaquero. Este sera bueno para cuidar los ganados y que ocupe el lugar que havia de tener un mozo de salario, no tiene delito particular cometido que castigarle. Su natural demuestra ser algo flojo y en cuanto a manas no dudo tendra aquellas que se producen en los de su pelo y sus obligaciones. . . ."

The comments of the mayordomos and the directives of the Estate officials suggest that, to them, society consisted of a set of fixed and hierarchical types, and that members of the upper social levels who had to supervise or control workers should assume, ideally, a tutelary and responsible relationship with them. The persistence of this view is exemplified in the work of Ruiz de Velasco (1937) where, in numerous places, he eulogizes various persons who displayed more or less ideal behavior toward their "gente" or local populace generally by providing some schooling for children or by avoiding the common practice of charging high prices at the plantation stores; celebrating these instances implies that many poor persons did not receive such treatment.

Religious teachings were a major source of the paternalistic attitude toward labor. All of at least the major plantations had a chapel, and at Atlacomulco all the resident personnel attended services together. For slaves, as for freedmen and Spaniards, all of the events of the life cycle important in Catholic practice received due attention: baptism, last rites, and Christian burial were routine, marriage within the Church was the ideal, and all the costs were borne by the plantation owners. The patron saint of the plantation was recognized by an annual fiesta, and so many other similar religious ceremonies were held that the mayordomo Urtado complained in a letter to the Governor in 1702 (L30 V56 E1), "Vmd remitira de avio doscientos pesos, que con las fiestas me matan esta gente."

Indian Labor at the Plantation

It was mentioned above that sixteenth- and seventeenth-century payments to Indian workers were listed in plantation accounts without much detail concerning the kinds of work they engaged in; instead, Indians were classified according to the five ways they were recruited. The account of 1588 contains a typical listing, including indios del cárcel, indios ordinarios (or de repartimiento) and extraordinarios, and indios naboríos. The list omits only the esclavos indios, whose importance had diminished greatly since the mid-century. The following paragraphs deal with the relative importance of the groups and the ways in which they became associated with the plantation.

Indios del cárcel were never very numerous: in 1584, for example, there were 13 Indians described as slaves at the mill, all from the Cuernavaca jail — "la gente condenada por el alcalde mayor." Probably they were kept in the plantation jail and in many cases were in irons but had enough freedom of movement to work. Most were obtained from the Alcalde Mayor of Cuernavaca on payment of an amount of money equal to their fine or debt, and worked off their debt to the plantation at the rate obtained by free Indian labor for fieldwork. Although few prisoners were employed in the seventeenth and eighteenth centuries at the Cortés mill, in the mid-eighteenth century prisoners were commonly assigned to trapiches and ingenios near Cuernavaca on the pretext, according to the Governor of the Estate, that the jails were not secure (L79, *Hojas Sueltas*). In response to local complaints, the Governor ordered in 1747 that officials were not to continue the practice, nor were plantation owners to receive prisoners unless they had been condemned to such service.

Indios esclavos numbered 165 in the inventory of 1549. Most of these Indians came from various parts of Mexico, but some were from Guatemala. Their total number was almost double the number of Negro slaves in 1549, making them an important source of labor, but it is possible that many worked in the obraje that was part of the complex of buildings at Tlaltenango; however, only 17 of the 83 Indian males were described as working in the oficio de la lana. Of the remaining males, at least 29 were assigned to the sugarworks, and nearly all had specialties: 5 were potters, 11 were cartdrivers, an equal number tended kettles in the boiling-house, and one was a smith. Another 37 males were not distinguished by occupation, nor were the 82 females listed as Indian slaves. There were 70 couples in the group, together with 21 single men and 22 single women, with no indication that any of the latter were children. Even after the abandonment of the obraje, some Indian slaves appeared in inventories, but their contribution to the total labor input was very small.

Indios ordinarios and extraordinarios performed the same tasks in the canefields, and the only major difference between them was the legal basis for their use: the ordinarios worked to satisfy the requirement that their villages or pueblos send a certain percentage of their males to work weekly for local Spaniards, and the extraordinarios were recruited to satisfy labor demands that could not be met by ordinarios. The earliest mention of the two kinds in the Tlaltenango records appears in the Rodríguez Acevedo contract of 1566 (L107 E3), in which it was specified that the Marqués was obliged to exert "facultad bastante para que podais sacar el servicio de indios ordinario y extraordinario que suelen dar y dan en los pueblos de mi Marques comarcanos al dho ingenio." Although this early reference suggests no practical difference between the two groups in respect of their obligation to satisfy the quotas imposed upon the pueblos, in fact the quota of extraordinarios had been only temporarily set by the Viceroy, whereas the contribution of ordinarios satisfied a permanent obligation that was subject only to adjustment of numbers.

Ordinarios were also called indios de semana, in reference to the fact that they worked from Monday until Saturday afternoon. After the workers returned home at the end of the week, a new group was sent the following week. The procedure was routine, with the same few neighboring villages and towns contributing to the plantation for many years. Usually, several pueblos contributed ordinarios simultaneously.

Ordinarios were occasionally referred to in the sixteenth and seventeenth centuries as indios de repartimiento, but the frequency of use of this phrase is very low compared to the frequency of use of the phrase Indios ordinarios. However, the use of the word "repartimiento" in this connection is important, since it is consistent with ordinary usage throughout New Spain and so serves as a reminder that the quotas were regulated by the Viceroy even in the Marquesado and on the haciendas of the Marqués himself. The number assigned from each pueblo was a percentage of the number of tributary Indians, so that when in 1590 the Viceroy Luis de Velasco allowed the use of 4 per cent of the 17,000 tributary Indians in the Cuernavaca region, the total available for work in any week was only 680. Only 69 ordinarios were at the time assigned to Tlaltenango, but the Marqués demanded 150 of the 680 for the plantation alone. In the end, 411 were granted by the Viceroy to the Marqués; 329 were from Cuernavaca, with only 24 allotted to the plantation, the rest going to the mines at Taxco (220) and Cuautla (85). Yautepec contributed 75 to this plantation exclusively, and Tepostlán 7, with the result that only 106 worked at the ingenio, in spite of the request of the Estate for more. Thus nearly two-thirds of the total number available within the jurisdiction of Cuernavaca worked for the Marqués. In spite of the fact that the local Indian population was still decreasing in 1590, the number allotted to the plantation represented a substantial increase over the 69 assigned by Viceregal order on 1 August 1575, when the number of ordinarios had been reduced by 21 from 90 (L282 E3).

Although this system of organizing a dependable labor supply continued with modifications for some time thereafter, it was in practice impossible to force the Indians into strict compliance. In 1592 (L245 E1) the mayordomo Ribaguda wrote to the Governor that the Indians of Yautepec had sent in one week only 61 Indians (the number expected was 75) plus 40 on account of arrears, and that they had promised to send each week as many as they could. The mayordomo told the Governor that if sufficient numbers were not forthcoming, he would repeat the latter's threat to send "them" to the mines, but there is no evidence that the threat was carried out or that it had the desired effect.

For ingenios, the quotas of ordinarios were abolished in the first decade of the seventeenth century, and the mayordomo noted in another letter (L263 E14) a short time later that the Governor found himself obliged to contract with Yautepec and Tepostlán to send 32 extraordinarios weekly as substitutes for ordinarios. He noted that the 32 workers generally appeared; sometimes there were more, sometimes fewer.

The plantation thus came to depend entirely on the Indian governors and principales, who bargained with the recogedor and "otras personas" who went to the pueblos seeking labor (ibid.). By the 1650's this method of labor procurement had become standard, as succinctly described by Blas Arias Tenorio in sworn testimony in a mid-century suit (L295 E137). However, seventeenth-century rental contracts contained clauses whose wording indicates that, even if the repartimiento system was no longer in effect, lessees of the mill expected a sustained flow of Indian workers. Nearly the same wording concerning the supply of Indians that had been used in both sixteenth-century leases was repeated in the Arias Tenorio contract of 1625 and the Castillo contract of 1682 (L269 E40; L30 V53 E1).

The hold of the governors and principales over labor appears to have become weaker by at least the early eighteenth century, and after the middle of the century there is very little mention in the Atlacomulco records of their importance. Two examples from the early eighteenth century show that, because the principales were no longer courted for labor, the system may have become less reliable than it had been, and that it had besides become necessary to rely on local royal and Estate officials for field labor. On 4 July 1705 (L30 V56 E1) the mayordomo Mateos wrote to the Governor of the Estate that the rains had stopped and a heavy growth of weeds required cutting, so that "con todo el auxilio al alcalde mayor [of Cuernavaca?] andamos al trote que todos handan con las Falegas de Pueblo en pueblo y pagandolos a 3 rr y no los allan." By contrast, in 1708 (ibid.) the same mayordomo wrote that if the plantation was to produce slightly more than 100 tons of sugar (18,000 panes) in the year, the Governor of the Estate would have to order the Indians of Ocotepec, Santa María, and Chamilpa to work at the plantation, and accounts of that year show that workers did come from these three villages as well as from Tejalpa and Tepostlán.

The labor supply must have become yet more uncertain in the years of disturbance that led to Independence. The mayordomo Marón Viñera wrote to the Board of Governors of the Estate in November 1818 (L219 E2) that it was impossible to persuade the Indians of nearby pueblos to work, and concluded by asking the Board to request from the Viceroy the "superiores providencias" necessary to obtain Indian workers, which the Board agreed to do, adding righteously that the Indians should desire to pay their debts by working at the plantation because "en el Ingenio siempre se les ha atendido y socurrido en sus necesidades." However, the threat of strife apparently increased so greatly and rapidly that only a short time later, in May 1823, the Board instructed the administrator to use only the gente of the plantation and to solicit Indians in the pueblos only in the case of the most extreme necessity (ibid.). This state of affairs represented a marked change from the conditions holding as recently as 1814 when, although the Ayuntamiento of Cuernavaca had refused to issue orders that the pueblos of Ocotepec, Tejalpa, Aguatepec, and Chamilpa send Indians to work at the plantation, the Governor did not regard it as impolitic for plantation officials to attempt to persuade Indians from those pueblos to work at the mill. These attempts had to depend more on persuasion than on any other means, as was suggested by the mayordomo when he wrote in the early nineteenth century that the Alcaldes Mayores Subdelegados had been accustomed to cooperate in obtaining Indian labor, "escitando de un modo suave y persuasibo a los Pueblos de la Comarca a que occurran al travajo de las Haciendas, pues estavan intimamente persuadidos de que al tiempo que alejavan la natural indolencia y ociosidad del Yndio, le proporcionavan su buen estar en la occupacion. . . ." (L27 V48 E11).

Irregularity and uncertainty continued to be characteristic features of the labor supply for the rest of the nineteenth century, continuing into the twentieth. The administrator Robles wrote to Juan Bautista Alamán on 13 April 1905 (L420 E15) that although workers had "always" been scarce at Atlacomulco, the situation had recently worsened. Administrators of other mills tried differing solutions: the manager of Temixco, for example, carried references from the President of the Republic to the Governor of Oaxaca in his search for workers; he distributed $2,000 in advances there, but few of the recipients showed up at Temixco and most of the advances were written off. Robles thought of searching for workers in Michoacán, because he had

heard in the early 1890's that workers there accepted low wages, and so he hoped that some might be persuaded to work at Atlacomulco (ibid.). Even the price of maize influenced the labor supply, leading to further uncertainty. In April of 1905 many persons arrived at the mill looking for work because of the high price of maize; Robles reasoned that if the price rose further many would leave to raise it for sale, but most were expected to leave in any case when the rains began in order to sow their plots. In sum, he concluded wearily, one day there were too many workers, the next not enough (ibid.).

Supervision and Payment of Ordinarios and Extraordinarios. In the sixteenth and seventeenth centuries, all the Indians were brought to the plantation in groups headed by an Indian alguacil, principal, or governor of the same pueblo, and in the eighteenth century by Indian capitanes who led workers of their own pueblos. The leaders were paid about the same as or a little more than the ordinarios themselves. They probably received some rations as well as very different treatment than did the workers, since their role in labor procurement was of such great importance. Probably most of the Spaniards who held supervisory positions at the plantation did not deal directly with Indian labor but rather passed their instructions through the Indian supervisors. There were too few Spaniards at the plantation, even in the late sixteenth century when their numbers were greatest, to have permitted direct supervision by them. However, Negro slaves, even women, were placed in charge of Indian work gangs.

It appears likely that most payments for extraordinarios went to the principales — in the period when they were important as far as the labor supply was concerned — who sent or brought them rather than to the workers themselves. The earliest indication that this mode of payment was customary appears in the receipts for payments in the years 1540–43 (L257 E12), where, for example, there is a statement that the governors of Amilpa were paid $70 for the work of three Indians who guarded the pastures of the Marqués. By contrast, the ordinarios were paid individually, and accounts and receipts place special emphasis on this fact; in the early years, a notary was usually present to witness the payment, but his presence ceased to be a matter of routine by the 1580's. The custom of paying principales, governors, or local priests for the work of extraordinarios is described in numerous accounts; for example, in that of 1589, it was stated that the money was given to the governors and priests "por junto para que los paguen de su mano e los hagan yr a travajar." In L263 E14, containing sworn testimony concerning conditions in the first decade of the seventeenth century, it was stated that extraordinarios were customarily paid through the governors.

After principales ceased to be important in supplying labor, payments were made to workers "en tabla y en la mesa," a phrase intended to show that each individual received from the hand of the mayordomo or his representative the amount owed. Payment was customarily weekly; indeed, such was the scrupulousness of at least the officials of the Estate in Madrid in regard to prompt payment that in 1789 (L298 E50) they ordered daily payment, but the system of weekly payment persisted.

Free Resident Labor

Naboríos were skilled and semiskilled laborers who were salaried and lived at the mill. As time passed, they came to be called la gente, dependientes, or criollos de Atlacomulco (1817; L219 E2); the change in name was accompanied by a change in their racial composition such that many of them, instead of being pure-blooded Indians, had Indian, Negro, and Spanish ancestry.

Tables 28 and 29 (p. 134) summarize the information about naboríos, showing that they were important in the work force from 1581 until at least 1593, and that later—particularly after 1618—their absolute contribution remained low, amounting only to the equivalent of somewhere between 25 and 30 full-time workers per year. Even before 1581, when detailed accounts began, their contribution had been important: in 1556 (L267 E26), for example, there were at Tlaltenango 150 Negro slaves and 50 indios naboríos.

The rates of pay given the various classes of labor for the years 1618–24 are only extrapolations of the data of the single and first year, 1618. The most important of the classes of labor are dealt with separately in the following paragraphs.

Carpenters were the highest paid: in the 1580's and 1590's, an Indian carpenter working full time might earn between $60 and $70 annually. Table 28 indicates that between 1581 and 1624 no more than 3 Indian carpenters were regarded as necessary, and in 14 of the years it was likely that the full-time equivalent of a carpenter or two would be hired. As was the case with other types of work, most of the persons hired worked only about half a year on the average.

The relatively low daily wage paid to smithy workers suggests that these naboríos either did very simple smithing or else were expected to help the master smith, a Spaniard. In only one year, 1583, were large numbers (13 full-time equivalents) hired, and in this year the daily wage was so low (4.6 granos) that we must infer their work to have been menial and simple. None were hired in the last 7 years of the period, when the plantation depended instead on hiring an itinerant master smith. It appears that in the 1580's and 1590's from 3 to 5 full-time workers in the smithy satisfied the needs of the plantation.

At the bottom of the pay scale with the irrigation tenders were the two groups of workers who labored within the mill itself, the potters and the purgery workers. The potters were engaged principally in producing the conical clay forms into which the moist sugar was placed and the jars on which these inverted cones rested. When the mill produced about 50 tons of sugar annually, probably about 10,000 each of forms and jars were needed, if we assume that they were used only once. About five full-time potters seem commonly to have been employed. Thus each might have produced about 40 forms or jars daily, which seems a very low daily output inasmuch as these rather simple containers did not require more than a crude finish. Even if it be assumed that one or more of the persons described as potters was engaged in preparing the clay and another in drying and storing the containers, it does not appear that the average daily output could have been burdensome. There appears to have been relatively little interannual fluctuation in the daily wage paid to potters, and apparently it nearly doubled sometime before 1602. As was the case with other naboríos, approximately twice as many individuals were hired as would have been needed if full-time work—300 days annually—had been customary.

The work in the purgery was light; boys of about ten were adequate for much of the labor involved in this part of the mill, and in fact were used there. At any time there might be one or two thousand forms being drained of molasses and requiring various simple operations. The light nature of this work was probably adequately expressed by the low pay, which nevertheless rose with other rates of pay between 1593 and 1602, and approximately doubled between 1602 and 1618.

Of the two categories of naborío field labor, that of laborer I cannot explain. Possibly laborers assisted the Spanish labrador. Both of these positions disappeared from accounts around 1600, lending support to this interpretation.

If the irrigation procedures described by Ruiz de Velasco and summarized above were in use in the period 1581–1625, then it is not very difficult to explain either the need for the irrigation workers or their low pay. Their job was very simple and required little physical labor: they were expected to make sure that the laterals from the main irrigation ditches remained open during the periods of irrigation. As the table shows, few were required except in the first few years after the plantation was returned to the Estate, and before that short time only one or two full-time ditch tenders were hired, probably because young male slaves were substituted for free Indian labor.

Transportation workers and ranch hands performed labor that requires little description. The oxen owned by the plantation were tended in most years by one or two Indians, together with Negro slaves. Many more Indian naboríos were hired to supplement the number of Negroes — even Negro boys were used — available for driving carts filled with cane from the fields to the mill. The muleteers drove mules laden with cane cut in fields that the carts could not reach, might have brought firewood to the mill, or carried sugar to the capital. After the 1580's, however, with nearly complete dependence on hired requas, very few Indians were hired as muleteers. I have included the work performed by the ranch hands at Mazatepec in Table 28, since much of the stock held temporarily at the ranch was destined for the plantation. As was the case with muleteers, fairly large numbers of Indians were hired as ranch hands in the 1580's, but after that time contributed only an insignificant amount of labor there.

Within the categories of miscellaneous and unassigned naboríos fall a small number of laborers receiving a low average wage, and whose importance decreased markedly in the period under review. Table 28 indicates clearly that only a very small percentage (6 per cent) of all the money paid to naboríos could not be assigned to clearly defined tasks.

Characteristics of Naborío Labor. For 19 of the approximately 45 years from 1581 to April of 1625, the names of the naboríos who worked for the plantation are found in the annual accounts. These years are 1581, 1583, 1584, 1587, 1589–90, 1592–93, 1602, and 1615–25; thus 12 of the 19 years are followed by a year for which the same data are available. For all the years except 1615–17 and 1619–25 total wage payments, months and days worked, rates of pay, and kind of work are given for each individual. It is thus tempting to try to describe some of the important characteristics of this part of the work force; I have listed the results in Table 30 (p. 135) for six of the most important occupational groups.

The results have numerous deficiencies owing to two major problems. In the first place, the gaps — particularly those from 1594 to 1601 and from 1603 to 1614 — may exceed the expectable length of time of association of most individuals with the plantation; very few persons working in 1593 or shortly before that date reappeared in 1602, and very few working in 1602 reappeared in 1615. And, of course, it is impossible to find any data concerning them before 1581 or after April 1625. In the second place, some confusion is introduced by the fact that a few surnames and Christian names were extremely common: for example, of the 122 purgery workers, approximately one-quarter (33) bore the surname Hernández. In some years, many of the new workers were called Miguel, in others Juan or Gaspar. This practice introduces bias toward increasing the average length of association with the plantation if identical names do not in fact represent the same person. In some cases, a little more certainty may be gained by inspecting the wage rates, which differed somewhat for the same class of work, or by some other identifying information such as place of birth or racial characteristics. In the case of the potters, of whom there were never very many at work at one time, I suppose it would require a remarkable coincidence to produce more or less identical sets of names on both sides of an interval without data unless these names represented the same individuals. In the cases of categories of labor containing more individuals, such as the purgery workers, such a coincidence may be more likely. The two major problems posed by the lists of names may thus introduce biases opposite in sign.

One might suppose another difficulty: that the same person may have performed different jobs at different times of the year, yet not be recognized as the same person because so few names were used by naboríos. Here, the column in Table 30 showing the average number of months worked per year is helpful. The range extends from four months to approximately seven, a relatively small difference, making it appear likely that most workers required about half the year free to raise subsistence crops. In the cases of the purgery workers, irrigators, and cartdrivers, not all could be employed all year because of the seasonal rhythm of plantation work. The only work that might have been nonseasonal was the potters', and in fact they do have the highest average in the column; even in their case, year-round production might have posed problems of storage that would make it undesirable, and I think that it was probably seasonal also. It appears likely, then, that most persons did work only about half a year, and in general only at one kind of work.

The categories "muleteers" and "Mazatepec" show both the shortest average association and the least time worked per year, in spite of the fact that both were paid relatively well; potters, by contrast, were among the lowest paid workers, yet their average length of service was notably longer. Perhaps the results for muledrivers and ranch hands were affected by the shortness of the records concerning these two categories. Muledrivers were not employed directly by the plantation after the early 1590's, and very few people were employed at Mazatepec after about the same time; in 1602, in fact, no one was employed at the ranch.

Of the purgery workers, slightly over half (67) of the individuals mentioned were identified as boys, or else could be identified as youths by reason of their low wages, about half the amount paid to an adult male. The boys raised the average length of time worked annually for the group as a whole, since their average was 7.1 months whereas the group average was slightly lower, 6.8 months. Of the boys 11 were identified as sons or other relatives of males working at the plantation, and it appears likely that they were an important source of labor for at least the purgery. Few boys were steadily employed in other naborío occupations except irrigation.

Later accounts contain much less information about free resident labor, but as the importance of Negro slaves declined in the late seventeenth century, resident workers became the most important source of more or less steady labor. In 1708 (L30 V56 E1), the mayordomo emphasized their contribution by writing that had it not been for "los Indios de Tlaltenango se hubiera atrasado esta hacienda en el todo."

Accounts of the eighteenth century did not distinguish naboríos from other labor, with the result that it is impossible to identify their contribution to the workload. The only information I have found concerning naborío labor in the mid-eighteenth century occurred as a result of their difficulties with lessees, who complained that they had not been fulfilling their obliga-

tions (L79 *Hojas Sueltas*). Their right to reside at the mill could be revoked by the Estate, whose Junta decided in 1823 to reduce the numbers of residents, keeping the best; at the same time it decided to get some good vecinos from nearby pueblos to reside at Atlacomulco (L219 E2).

Near the end of the eighteenth century, when there were virtually no slaves at Atlacomulco, its population consisted of 230 persons, of whom 72 were classified as vecinos (1790; Mazari, p. 151). About 20 years after Haven had reported in the 1870's (p. 243) that "Indian" workmen lived in "cane-huts" just outside the mill, Velasco (1890, p. 56) stated that Atlacomulco had a population of 398. The relatively small size of this group may be judged by contrast with the population of 1,270 at Temixco; its resident population, like that of Atlacomulco, had approximately doubled within a century (Mazari, p. 156). By the 1820's, a resident work force of about 150 was considered adequate to produce 500 tons of sugar annually — slightly more than Atlacomulco produced — so probably the number of dependientes at the plantation was not quite sufficient (Ward, Vol. 1, p. 68).

Debt Peonage. All classes of labor at the plantation, including Spaniards, received wage advances, but probably this method of attracting labor applied primarily to naboríos; in 1586, for example, when naboríos owed advances of $572, four Spaniards owed only $73. However, ordinarios and extraordinarios received wage advances as well, and a mid-eighteenth-century listing of debts shows that both nonresident labor and their capitanes had received advances (L230 E9).

The advance as a method of recruiting labor was not very successful, and there is no way to reach conclusions from the records concerning its effectiveness in keeping workers at the plantation. In the eighteenth century, such debts were written off from time to time. The ineffectiveness of wage advances became especially noticeable during the Revolution; the administrator wrote to the Junta on 25 November 1822 that the debts of the Indians of neighboring pueblos were insufficient incentive to bring them to work, and in the same year he advanced money to some but they failed to appear for work (L219 E2). A mayordomo must frequently have had to ask himself whether the size of an individual's debt was sufficient to justify the cost of sending one or more representatives of local justice to bring the debtor to work, and whether there should be an upper limit to the numbers of debtors and the sizes of their debts, particularly since under such conditions they may not have worked very efficiently.

The haciendas customarily maintained stores in the eighteenth and nineteenth centuries where workers could buy cloth and food. These were important sources of income for the plantations, and workers could run up debts at the stores in addition to receiving wage advances. Alamán described their operation in the middle of the nineteenth century (*Documentos*, Vol. 4, p. 55), stating that it was customary to pay workers by giving them one-half their wages in cash and the other half in vouchers to be used for purchases at the plantation store. Not only did this practice leave mayordomos free to charge the prices they wished — at Atlacomulco earlier in the century the mayordomo was even allowed to run the store on his own account and for his own profit — but it also removed the burden of obtaining each week half the cash requirements for labor. This may have been a boon to the owners and managers, but Alamán recognized it to have been an onerous burden on the workers, and decided in 1850 to pay them only in cash. The response was immediate and gratifying — so many persons applied for work at Atlacomulco that it was possible to lower the wage rate at the plantation below that of other haciendas.

Treatment of Indian Labor. Indian labor was regulated by viceregal orders based on a conception of Spanish rule that was more closely identified with humanitarian ideals than with economic change or a thorough reworking of the structure of Indian society. I cannot discuss the gap that must have existed between local practice by entrepreneurs and the humanitarian ideals of the government, but many of the viceregal orders had the intended effect of narrowing the gap. Their purpose is shown in the collection of documents edited by Silvio Zavala, and scattered references to their effects are contained in accounts and other documents pertaining to Atlacomulco. For the most part, these orders concentrated on regulating the kind and amount of work performed by Indians, the rates of pay, and the prohibition of forced labor. The most important is the well-known order of 10 November 1599 (L258 E9) specifying that Indians might be used for fieldwork in the plantations, but not in the mill or boiling-house, where conditions were regarded as dangerous and unhealthful. Thus, in almost every inventory one

or more Negro slaves were described as lacking an arm from having caught it between the rollers of the mill, which must have been difficult to stop quickly. Only a very few — possibly one or two of the thousand associated with the mill throughout its history — died as a result of falling into a kettle of boiling juice, but the high temperature and humidity of the boiling-house, where nearly constant movement must have been required, were unhealthful.

The order also specified that work was to be performed only voluntarily, but it seems doubtful that this part of the order affected the recruitment of Indian labor for Tlaltenango because the system of procurement through principales continued after it was issued. Furthermore, the major sixteenth-century change in Indian labor — the reduction in the numbers of Indian slaves living on the plantation — occurred long before the order was issued. Nor did the prohibition on the use of Indian labor in mill and boiling-house affect Tlaltenango greatly, since Negroes had supplied most of the labor in both places for several decades.

Enforcement of the new edict was placed in the hands of a person designated Juez de Ingenios, and the twelve mills of Cuernavaca, Yecapixtla, and Cuautla districts were assigned to one person. He received a salary approximately equal to that of the mayordomo of Tlaltenango, lived at the mills, and received payment from each according to its production. The highest contribution from a single unit was from Tlaltenango, which paid $250, so that it must have been the largest mill of the dozen.

Estate officials, particularly those who viewed the labor supply with the advantage of distance in Madrid, did not rely solely on Crown edicts to guide them in the treatment of Indian labor. In 1789 the following statement arrived from the Dirección General del Estado in Madrid: "Deve cuidar el Mayordomo, que las faenas, que se ofrecen en la Hacienda no sean excesivas pues en pasando de dos horas son irregulares, y por consiguiente se grana la conciencia, y si esta ignorante el Dueño, ô quien le representa, deve saverse es en perjuicio de la Hacienda, aunque digan se adelanta, a demas de que Dios no quiere se sirva nadie de trabajo ageno sin la correspondiente paga, y mas siendo hecho por unos pobres infelices, como son tales gentes. . . ." (L298 E50).

IX

Costs and Productivity of Labor

THIS chapter deals first with the costs of labor and then describes changes in its productivity. The first section is further divided into two parts: the first deals with the value of rations received by the resident labor of the plantation, and the second with wages paid and the costs of buying Negro slaves. The costs of labor, as might be expected, dominate the plantation accounts; by contrast, land was cheap, and repairs to buildings and equipment, although high in some years, amounted to relatively little when prorated over a number of years.

Rations

Only three classes of labor received rations: skilled or supervisory workers, Negro slaves, and naboríos in the sixteenth and seventeenth centuries.

Mutton. For Spaniards, mutton was important in the diet, as shown by the fact that their ration was half a carcass per week, as it had been since at least the 1550's at Axomulco (L282 E4). Perhaps the weight of half a carcass was 15 to 20 pounds; unlike the beef ration, the mutton was not weighed.

In the 9 years for which data are available between 1596 and 1603 the annual costs of purchasing sheep were relatively high, ranging from a low of $392 in 1597 to a high of $869 in 1596; seven of the annual values ex-

ceed $600. By contrast, for 9 years between 1769 and 1822 the range went from zero to $684 in 1815, the only year when purchases exceeded $600. Annual consumption in the late sixteenth century was within the range of 500 to 700 sheep, at a time when Spaniards worked at the mill a total of approximately 3,500–4,000 days per year; since the midpoints of these ranges are in the ratio of approximately 1:6 (one sheep per six days worked), or approximately one sheep per Spaniard and week, obviously mutton was dispensed to other persons at this time. Some went to convents and some to Indian principales (1595; L262 E5 D7), so this mutton should be charged to the expense of procuring Indian labor, but it is not possible to do so with exactness because of lack of information: some sheep were lost, as the records indicate; some went to the plantation dispensary for inpatients, and some — probably not very many — went to Negro slaves. I have assigned the difference between total consumption on the one hand and rations and gifts on the other to the costs of Negro labor.

For the years 1619–24 for which data on consumption and time worked by Spaniards are available, it is obvious that the ratio of sheep consumed to days worked by Spaniards had changed considerably, from the earlier 1:6 to more than 1:10 in every case except 1624, which means that rationing of mutton to persons other than Spaniards had been much reduced. For the years following 1778, data concerning consumption are not available, but the numbers bought may be distributed throughout the year at the rate of two to four weekly; this implies that four to eight Spaniards or other persons entitled to such rations were employed at the plantation — figures in line with the actual numbers of supervisory personnel. Thus, it seems fairly clear that other persons could not have received more than an insignificant amount of mutton, and so I have charged the mutton entirely to administrative and supervisory personnel.

I have estimated the contribution of the mutton ration to Spanish wages by first dividing the total number of days worked by all Spaniards by twelve (the number of days worked in a two-week period for which one whole sheep was received) to obtain the approximate total consumed as Spanish rations. The value of the sheep was then divided by the number of days worked by Spaniards to obtain the addition to the daily wage from the sheep ration. Before 1600 the daily addition was small, amounting to slightly less than one real, or nearly equivalent to the daily wage of an Indian at this time. Since the price of sheep remained steady in the 1620's, the contribution to daily wages then was relatively higher. Probably it was near this level in the eighteenth century, when the numbers bought in various years changed little from the numbers bought after about 1615.

Beef. Spaniards received 20 pounds of beef weekly, adult Negroes and naboríos 10 pounds each in the late sixteenth century; children, at that time not a very high percentage of the slave population, were issued half the adult ration. In the late sixteenth century probably between nine and fourteen head of cattle were slaughtered weekly (as in 1587), and it appears that a surplus was available for sale, probably to nonresident Indian labor. Since beef was not eaten during Lent, the maximum annual issue to a single naborío was about 450 pounds, and the account of 1581 shows that the maximum issued to a single Spaniard, in this case Padre Vicario Juan de Ayllón, was exactly twice that much. The issue of beef to Negroes, although supposed to be the same as the naborío ration, resulted in different annual totals: in 1584, for example, adult Negroes received a maximum of only 400 pounds in the entire year.

The difficulty in allocating costs of the beef rations lies in the fact that the dressed weight of cattle is nowhere revealed in the accounts. In the effort to establish the average weight of range cattle used for rations, I am going to assume that the customary rations were in fact given; the weights stated to have been issued need not be questioned, since there was always a balance used exclusively to weigh meat at the plantation.

There are two sets of evidence from the late sixteenth century that suggest the average dressed weight to have been very low. The first evidence is found in the account of 1587, which gives the number of cattle killed each day when slaughtering occurred, usually Saturday and Tuesday. There was a sudden rise in the number killed in the last week of August, and this larger weekly number continued until the end of the year; a note explains that the increase was due to the arrival of slaves from Tuxtla. The document L262 E2 D3 contains a list of these slaves, clearly identified as those brought from Tuxtla in August of 1587; there were 44 adults and 4 children 10 years of age or younger, so the weekly beef ration of these 48 persons should have been 460 pounds. Until they arrived in August, 9.2 head of cattle had been slaughtered weekly for 28 weeks; after their arrival, the average number slaughtered weekly was 13.5 for 16 weeks until the end of the year, or 4.2 cattle for persons whose normal total ration should have been 460 pounds. Dividing this requirement by 4.2 yields the very low average dressed weight of approximately 110 pounds.

Another set of data yields a slightly higher value. For the six consecutive years 1597–1602 the accounts give the number of cattle slaughtered annually for rations, and I have calculated for the same years the full-time equivalent of the Spanish and Indian naborío labor in order to estimate the pounds of beef consumed per year by these workers, using the figure of 900 pounds per Spaniard annually and 450 pounds per naborío. The Negroes were rationed at the same rate as the naboríos; possibly the Negro rations result in a slight overestimate, because I have not distinguished children from adults. Dividing the total pounds of beef consumed by the number of cattle slaughtered yields remarkably consistent values of average dressed weight, which range for the six years between approximately 140 and 165 pounds. The average live weight might have been 280–330 pounds. Although the figures are low, it should be remembered that many of the animals consumed were described as novillos — young males — as well as toros, and that if instructions (Appendix F) had been followed, some were females of various ages.

The discontinuity between sixteenth- and early seventeenth-century practices is particularly striking in the case of beef rations. From annual consumption of over 600 beeves, consumption dropped to fewer than 200 by 1619, and in 1621 and 1623 fell below 100. Consumption of about 100 head annually, including a small number of overage oxen, seems to have been normal until at least 1769, after which date only the total value of the meat ration was given. The approximate average annual outlay for 1779, 1780, and 1782 was $350; for

1785, $504; and for 1828, $196. By contrast, before the end of the first decade of the seventeenth century, the annual cost was well over $1,000, and in 1600 alone it amounted to $3,294.

Corn. As was the case with other rations, until about 1615 consumption of corn per capita continued at a high rate among all three groups of workers, Spaniards, Indian naboríos, and Negro and Indian slaves. After about 1615 the rations of all three groups were reduced by approximately one-third, and total consumption halved until at least 1624. Total purchases demonstrate the pattern, remaining over $1,000 for the years 1587–1601 and reaching a high of $3,500 in 1597. In the 1620's annual costs remained below $1,000, not exceeding that figure until 1811 when the price rose to $2-5-0 per fanega and 1819 when the price was $4-6-8; however, few data on purchases and consumption are available for the eighteenth century.

Two circumstances are associated with the smaller consumption that began in the early seventeenth century: the high price of corn and a smaller resident work force at the plantation. It appears likely that, after Negro slaves ceased to be important at Atlacomulco sometime in the eighteenth century, very little corn was issued as rations; thus when there were no Negro slaves at all (1831), total annual consumption was only 324 fanegas, and about 80 per cent of this amount was fed to mules and horses. The latter had always required some corn to supplement the grazing available.

Reconstruction of the sequence of events that led to the very great differences between the period before about 1615 and the time following has to be based on incomplete data, inasmuch as there are no corn ration lists after 1619, nor even figures showing annual consumption except for 1831, more than two centuries later. However, with the use of the available data a reasonably satisfactory reconstruction is possible. The graph showing prices of commodities bought and sold by the plantation (Fig. 3, p. 19) indicates that although sugar prices seem to have peaked in the 1590's and began to decline after about 1600, the prices of supplies such as corn and mutton did not peak until sometime in the 1620's. Corn was not purchased in large amounts by the plantation, but was instead supplied from tributes — much from Cuernavaca — at an assigned value of $1 per fanega, beginning in the 1580's. However, the Governor's accounts beginning in 1596 show that much money may have been lost to the Estate by this transfer, since the sale price in Mexico City

was often twice the assigned value, and I have used the sale price of large lots of Estate corn for the years when it was available rather than the assigned value of $1. It must have become apparent in the 1590's that the price of corn was rising more than that of sugar, since the annual consumption in the 1590's was much less than that in 1587 and 1588 and probably 1584 (total consumption is not available for 1584, but it must have been in excess of 2,000 fanegas). Even by the 1580's, monthly rations had been reduced below what they had been in 1542 at Axomulco.

It seems likely that the reduction in the corn ration was due to its rising price, and the latter must have depended in part on the depopulation that was so important in the late sixteenth century in New Spain. Local corn was thus in short supply from time to time after 1600. In 1608 only a third of the tribute corn received by the plantation was from Cuernavaca, and the rest came from places as far away as Toluca and Chalco; the reason given for reliance on supplies from places so distant as to result in high freight costs was the lack of corn in Cuernavaca. About a third came from Toluca for the periods 1609–11 and 1613–15, and dependence on Toluca continued until the accounts end in 1625. Freight costs were not always disadvantageous, for the account of 1619 contains a notice that the maize bought that year in Toluca, including freight, cost $2-4-0, the same as in Cuernavaca itself.

Instead of relying wholly on purchasing corn and using tribute corn for rations, the plantation began to grow its own; the first indication that this possibility was being explored dates from 1621, when 153 fanegas were harvested from the milpa, yielding about one-fourth of the total consumed. By at least the early eighteenth century, the gente at Tlaltenango were allowed to use nearby land for subsistence crops, including corn, and it is possible that this arrangement began in the 1640's when the mill was moved to Atlacomulco.

The inventories provide little information concerning plots of maize in the eighteenth and nineteenth centuries, except to show that the plots were so small that the plantation could not expect to meet its requirements by growing its own.

Wheat and Chocolate. In addition to maize, Spaniards received a generous ration of wheat in the sixteenth century that was commuted to a cash allowance sometime in the 1580's. The Axomulco records of 1541–42 show that most Spanish employees received a fanega of wheat monthly, whereas the mayordomo and sugar-

95

master received twice as much. Slightly more than a decade later the Spanish mayordomo and another Spanish official of Axomulco received 2 quintales monthly, valued at $2 per quintal. In the four-year period of this account, the value of the wheat ration for these two employees alone was $384. By 1585 (L257 E13 D4), Spaniards received $2 monthly for "pan," rather than a ration of wheat, raised by 1590 to $3 for most and to $4 for some. The rates remained the same until at least 1602 when most received $3 and the Capellán twice as much. The custom had altered considerably by 1644, when there were only 4 Spaniards at the mill — the allowance was then called "pan y chocolate," and amounted to $10 monthly for the purgador, slightly more for other officials. The chocolate ration continued through the eighteenth century and was $5-5-0 monthly for all eligible employees in 1768, when it was more fully identified as consisting of cacao, sugar, and cinnamon. It was issued to the mayordomo and field supervisors (caporales).

The Mayordomo's Table. An important supplement to the wages of the administrator of the plantation was a food allowance that at times equaled or exceeded his salary. Probably most of the administrators were married and had servants, so the allowance should be interpreted as the support of varying numbers of people directly dependent upon them. In addition, unmarried Spaniards working at the plantation and skilled help working there temporarily — smiths and master carpenters — ate at the mayordomo's table. The allowance also included the costs of food for special occasions, as in 1595, when the Governor, the Juez Privativo of the Estate, and other dignitaries attended the inventory of that year. For most of the years from 1588 to 1608 there are notes in the accounts that the totals included wine and fish at Lent for Indian governors and principales who persuaded Indians of their pueblos to work at the plantation, as well as various Easter gifts to these and other local persons with some control over the labor supply.

The amount of money spent in this way caused the controller and Governor to question the totals from time to time, and in 1588 the administrator was ordered not to exceed $300 annually on this account (Appendix E, p. 115). This order was not obeyed, but after administration of the mill was resumed by the Estate in the mid-seventeenth century the amount spent on the mayordomo's food and entertainment was much less: for example, for the five months from August to De-

cember in 1645 he spent at the annual rate of approximately $375, at the same time as he received a lower salary than was customary earlier in the century. In the early eighteenth century his salary dropped still more, to half the amount it had been in 1620, and continued at a low level for the rest of the century; it appears from the large amounts listed under "gastos de casa" in the late eighteenth-century accounts — in 1782 nearly four times his salary — that he could compensate for the low salary by spending rather freely on the table. Although the salary was doubled sometime after the turn of the century, table expenses continued at a relatively high level until the former was increased to $1,500 between 1832 and 1847 and table expenses were reduced to a level near that of the earlier part of the nineteenth century. The data I have found show that table expenses were always at least half the value of the administrator's salary, and that in many years — particularly in the late eighteenth century — they were more than double his cash income. In the summary of labor costs I have treated these expenses as part of the costs of administration.

Clothing Rations. As was the case with many other rations, Negro and Indian slaves and naboríos were treated similarly in clothing rations, since both groups were given an annual ration of the coarse woolen cloth called sayal, usually at Christmas. At times this was commuted to cash, as at Tlaltenango as early as 1549 (L257 E14 D183). The Axomulco account of 1553–56 stated that 10 jaquetas were given to 10 naboríos, each consisting of 3 varas of sayal and valued at $0-4-0, for a total of $15; the standard ration was stated to be 1 jaqueta per year. The rations of sayal to naboríos varied from time to time, being 4 varas in 1581 and 1584, but only 3 varas in 1583 and 1595. It is possible that the ration of sayal to naboríos was eliminated about 1615, since the total amount bought annually after that date was much less than it had been, and it seems fairly clear that after the Estate resumed the administration of the plantation in 1644–45, naboríos no longer received a sayal ration.

Negro slaves received 5 varas of sayal as an annual ration, although in some years they must have received less. In addition to the cloth, each Negro was supposed to receive at least one frezada annually and in the early years mantas were distributed to them as well. Indian slaves apparently received neither of these in the sixteenth century.

It has been possible to distribute fairly exactly the

costs of cloth between Negroes and naboríos because the prices can be calculated for most years. In cases where the accounts do not specify the amounts issued to the two groups, it is reasonable to assume that each slave received 5 varas and to assign the remainder to the naboríos.

The amounts spent annually on clothing did not amount to a great deal in the sixteenth century, ranging in the years for which I have data from $301 (1595) to $797 (1600). The costs, like the costs of other rations, were much reduced in the early seventeenth century. The naboríos suffered most by the reduction, since it was achieved by eliminating their rations. Their rates of pay changed little after about 1610, and the clothing ration had meant annually a peso or more in value.

Wine Ration. This ration excited more comment than any other of the sixteenth century. From time to time the controller questioned the rather large sums spent on wine. For 1587–89 the amounts spent were, respectively, $305, $269, and $444; in the last year 58 arrobas were purchased. By 1598 the amount had been reduced to $31, and in 1603 it was only slightly more, $41. Obviously the 1580's stand out as the years of lavish consumption of wine. The amounts allowed the plantation fell off sharply in succeeding years to modest quantities that were claimed to be mostly for medicinal purposes and important guests, as in the account of 1598. In the years when large amounts were spent on wine the totals were equivalent to the cash salary of an experienced Spanish worker.

Lenten Rations of Pulses and Fish. The first of these two categories consisted of habas, beans, garbanzos, and lentils, of which the latter two were of little importance; by bulk, the habas were the most important. It is possible that some of all four kinds were eaten at other times of the year, but there is no information available on this point, and many of the individual annual accounts are explicit only concerning their consumption during Lent. The changes in the price of habas were similar to changes in prices of other commodities, rising in the sixteenth century, lower and changing less in the seventeenth and eighteenth centuries. I have not continued the analysis beyond 1747 because after this date there were too few slaves to result in high consumption of these minor items.

Other Rations. The rationing of tobacco to the Negro slaves was first indicated in the account of 1644–45, when a tercio was bought for them at a cost of $21. An-other tercio was purchased in 1707–8, and in 1718–19 the slaves were expected to content themselves with tobacco of lower quality, since they were given a tercio of "tobaco sacate" costing $8.

Rations in the Eighteenth-Century Accounts. Sometime after the middle of the eighteenth century different categories were used to account for rations, one being "rations for meseros." This class of meseros, persons working for a monthly wage rather than by the day or week, may be taken to include supervisory and skilled workers. In 1768, the first year available in which this category appeared, the sum of $826-4-0 was spent on such rations, and the entry included the information that both free workers' and slaves' rations were included in this amount, which nevertheless excluded meat and maize. In the following year the amount was $946, but in 10 years between 1779 and 1831 for which the figure is available, it remained below $665.

Summary. For many years in the late sixteenth and early seventeenth centuries on the one hand, and in the late eighteenth and early nineteenth on the other, the values of most of the items issued as rations are available. However, in view of the absence of complete information for many of these years, I have restricted comparison to 5 years of the earlier period (1596–1600) and to 5 others in the later (1768, 1779, 1780, 1782, and 1785) when the values of all rations were stated. In the late sixteenth century, the average annual cost of all rations was $5,947, whereas in the late eighteenth century the average value was approximately half that amount, $3,049. The partial data of many other years support the inference to be drawn from this comparison, that the reduction was a permanent feature of mill accounts begun in the early seventeenth century partly in an effort to reduce costs and partly because of a loss of workers. Although the production of the mill increased in the late eighteenth century, it was no longer necessary to issue rations to many workers because nearly all were free and no longer expected them.

Costs of Negro Slave Labor

These costs fell into two classes, the costs of purchasing slaves and their maintenance. The purchase prices of nearly all the slaves bought for the plantation are available, but the costs of their maintenance are somewhat more difficult to establish.

Values of Slaves. Figure 18 (p. 83) based almost entirely on inventory evaluations, shows for various dates the complicated dependence of value on age, skill, sex, health, and racial composition. I have selected these dates as adequately representing other years when inventory values were stated.

Age emerges as the most obvious factor in determining value, with the data showing that from 1596 to 1754 the value of a slave increased from birth until about the age of 25, after which it leveled off; aging did not operate so consistently to depress value as did maturation to increase it, but the data — particularly of the 1740's — suggest that after about the age of 45 or 50 value began to decrease. Thus the 20 years from age 25 to 45 were regarded as the most productive in the life of a slave. Skills of males were also important, and in cases where these were described, the sugarmaster was usually the most valuable slave, as might be expected: no other skill possessed by a worker so greatly or directly affected the profitability of a plantation. In general, males were more valuable than females, but in the years near 1600, when it seems to have been hoped that natural increase would supply at least part of the slave labor demanded by the viceregal edicts and therefore the value of females was enhanced, the most valuable slaves were two females valued at $500 each. Chronic or permanent physical disability operated to depress the value of a slave, and the variations in disabilities were so great as to provide the major source of scatter on the diagram. It seems likely that at a given age mulattoes had less value than at least Negroes born in Africa, but since the Atlacomulco data deal with two so markedly different populations — an early one dominated by African-born and later ones dominated by mulattoes — no certainty can be attached to this conclusion.

Owing to the effect on prices of slaves of the factors dealt with above, I do not regard it as worthwhile to present more than general statements concerning the changes in prices of slaves. The average value at maturity, shown by the dashed horizontal lines in Figure 17, is adequate to the purpose. The value of mature Atlacomulco slaves peaked in the late sixteenth and early seventeenth centuries, but by 1655 had reached a level it maintained with little variation for more than a century; purchase of the Atotonilco slaves in 1721, some of whom had children, probably helped to maintain the average value for a while. The effect did not persist, however; within a decade the value of a mature slave had sunk by $100 to a low level that reflected a growing indifference to the maintenance of this class of labor. The decline in value between the decades about 1600

to the year 1655 indicates not only the effect of the many factors discussed above, but probably also the partial return of Indian labor to the service of the plantation and possibly a change in the value of the peso. The slaves present in 1754 were nearly all mulattoes, mostly immature or disabled, and few gave long service to Atlacomulco.

In order to describe annual costs of slave labor, I have adopted the procedure of prorating over a number of years the purchase price of slaves, since most were bought in lots rather than singly and this practice had the effect of causing some annual costs to be extremely high in contrast to those of other years when none were bought. To reduce calculations, I have distributed the purchase price of slaves over the average working life of slaves appropriate to the decade in which they were born, adding this annual cost to the cost of rations. Since, in general, the course of amortization followed fairly closely the total number of slaves, I have not presented it in graphic or other detailed form, merely adding it to the total annual costs of slave labor. The results (Table 31, p. 135) show that the total costs of maintaining and buying Negroes, heavy in the 1580's and 1590's, owing to numerous purchases, were greater from 1583 to 1600 than the costs of all Indian labor but that from 1621 to 1624 the costs of both kinds were more nearly equal.

Relative Efficiency of Negro and Indian Labor. Whether or not imported Negro slaves and their descendents worked harder and more efficiently than did Indians is a question with a long history and some relevance to this discussion. If these differences were real and large, they may mean that Negro labor was in fact cheaper than Indian when the daily cost of Negro labor was about the same as that of Indian, as in the 1620's.

The only extensive comparison that I have found describes conditions in the 1650's. It consists of sworn testimony submitted by witnesses supporting the Governor Diego Valles in a suit he brought against the Estate for the labor of 53 Negro slaves missing at the time he assumed responsibility for the plantation. Since Valles was then forced to rely heavily on Indian labor and demanded compensation for his expenses, the testimony centered on the wages that might be earned by Negroes in order to fix the amount owed Valles by the Estate. Not surprisingly, Valles' witnesses agreed that the Negroes deserved a higher wage than did Indians, but since many of the witnesses were owners or managers of haciendas their testimony deserves consideration.

The lawyers of the Marqués did not contest their testimony, and it seems reasonable to assume from their statements that a difference in productivity did exist.

The arguments of the witnesses may be reduced to a number of points on which there was general agreement. Negroes were claimed to work longer hours than did Indians, from 3 or 4 A.M. to 10 or 11 P.M., whereas Indians worked from 8 or 10 A.M. until 4 or 5 P.M. Indians would not work at night, but Negroes would; since night work was done only in the mill where Indians were not allowed to work at all, there can be no disagreement concerning at least this point. Indians required more supervision than did Negroes: when they finished an assignment, they did nothing else if not directed, whereas Negro slaves worked continuously. Someone had even to bring the Indians to work. The quality of Indian work was poor: "El lavorio de los naturales en semejantes haciendas es de ninguna consideracion"; but we have seen that the Cortés mill could not have continued without it at any time before the nineteenth century. Negro slaves were capable of becoming skilled workers — smiths, cartwrights, and sugarmasters — and many did so at Atlacomulco. It is true that fewer Indians assumed these positions of responsibility.

The ratio of value of Negro labor to that of Indians was stated as 2:1, 3:1, and 4:1, with even higher relative value placed on the work of skilled Negro workers. However, the difference in wages actually received, or that should have been received, at the time was stated merely to be at least 4:3 or 5:3, much lower than the difference in actual wages had been a century before, when Negro slaves used at Axomulco were paid for at the rate of $3 monthly, and Indians doing the same work received only $1 (L277 E4). If we accept these statements and ratios as expressive of actual differences, the small difference between the costs of Negro and Indian labor in the period 1581–1624 means that the former was indeed more profitable. The comments that I have summarized from the mid-1650's show that this advantage persisted until at least that date. Perhaps the apparent difference did not persist into the eighteenth century, because Estate officials did not make strong efforts to maintain Negroes as even the skilled core of the labor force then.

Total Costs and Inputs of All Kinds of Labor

The reality and impact of the viceregal edicts regulating Indian labor, partly effective in 1599 and fully effective on 31 December 1600, may be judged by comparing the annual total of days worked for several years before 1601 with the total in 1601 (Table 33, p. 135): for the years 1596–98, annual totals were slightly in excess of 34,000; in 1599, the total fell in response to the first of the edicts; in 1600 more than 34,000 days were worked by nonresident Indians, but in 1601 their contribution was almost halved, and in 1602 it fell again. A less dramatic change occurred with respect to naboríos, whose contribution declined by about 25 per cent from 1600 to 1602. In the period 1620–24 their average annual contribution was 8,254 days, fairly representative of the range of the 5 years. Although the need for supervision had diminished owing to the presence of fewer workers, the Spanish contribution rose from 1600 to 1602, but then fell irregularly, reaching the low figure of 806 days by 1612.

The averages in Table 31 show dramatically the contrasts between the first two periods. The average naborío contribution was halved, the contribution of nonresident Indians fell sharply to one-seventh its earlier size, and the input of all Indian labor diminished to about one-quarter of the annual average of the last two decades of the century, showing by its change the earlier dominance of extraordinarios and ordinarios in the Indian labor supply. Spanish labor fell to 60 per cent of its earlier size, whereas the average number of days worked annually by Negroes changed little. Of course, it had been the intent of the edicts to substitute Negro for Indian labor, and apparently Estate officials had tried to conform to the intent of the edicts in spite of the high cost of slaves and their high mortality in the first few decades of the century. However, the limitations on increasing the numbers of Negroes were powerful enough to require employment of at least some Indian labor. Under the circumstances, there was no way of avoiding a marked drop in the total annual input, with the result that the average number of days worked annually from 1621 to 1624 was slightly less than half the average annual input of the decades before 1600.

The total Indian wage bill reflected these changes. From $5,689 in 1600, it fell to $2,928 by 1605, and between 1617 and 1624 ranged from $1,333 to $2,586. The abolishment in 1599 of the repartimiento for sugar plantations had been accompanied by orders emphasizing that the daily pay was to be one real, food was to be supplied, and payments for travel made at the rate of one real for six leagues (Zavala, Vol. 4, p. 358). Table 31 thus registers the effects of the edicts, showing that the daily wage of a nonresident Indian laborer was twice as high in the period 1621–24 as it had in 1583–1600. Although individual payments to naboríos varied more widely than did payments to nonresident Indians, their average daily cost dropped slightly sometime after 1600 because of elimination of rations; their average daily wage doubled (Table 29). Not only were fewer Spaniards required at the mill, but those present received in wages and rations only a little more than half the amount they had received before 1600. The daily cost of maintaining a Negro slave rose somewhat between the two periods. In general, because of the decreased supply of Indian labor, total annual labor costs were nearly halved, suffering less reduction than did the total numbers of days worked because the cost of nonresident Indian labor had doubled.

After the well-documented period 1620–24 there next appear data of two years from 11 January 1644 to 31 December 1645. By this time, the administration had abandoned the attempt to increase or even maintain the numbers of Negro slaves in the face of high mortality rates, so Negro labor contributions in these two years were only 10,500 and 9,800 days, about half the annual average two decades before. The contribution of naboríos remained very near the average for 1620–24 (8,254): 8,967 and 8,682, about equal to the input of Negro labor. Similarly, the Spaniards' contribution changed little, remaining approximately the full-time equivalent of three workers. What had changed greatly was the amount of work done by nonresident Indians, no longer called ordinarios or extraordinarios, but simply Indios: in 1644 they contributed three times as many days (10,095) as they had on the average in the period 1620–24, and in 1645 five times as many (25,800). The total number of days worked in 1644 (36,612) thus remained very near the average for the years 1620–24, but in 1645 increased to 45,251 — attributable entirely to the great increase in the number of days worked by nonresident Indians, since all other categories experienced slight declines from 1644 to 1645. It appears, then, that partial recovery had been made during these years toward the high labor inputs of the 1580's and 1590's.

Although the data of accounts for the next two years in which they are available, 21 March 1706 to 10 March 1708 (L30 V58 E1), are not presented in sufficient detail to enable fixing the total days worked in these two years, the total labor costs — less the administrator's table — were at least $10,954 in the first year and $11,-

173 in the second. By this time, therefore, the total annual costs were approximately $2,000 greater than they had been in 1620–24, and we may infer larger labor inputs in the absence of a rise in wages. Accounts dealing with the three years from 20 August 1718 to 20 August 1721 also lack detail, but contain enough information to fix minimum labor costs at $17,728, $16,139, and $17,870, respectively. The average annual cost was thus near $17,000, an increase of about $6,000 annually over the previous decade and nearer the average of $25,107 for the years 1780–1822.

The existence of flexibility in the labor supply after the middle of the eighteenth century is demonstrated by the contrasting data of inputs in 1768 and 1769. The total cost of labor in 1768, excluding the cost of maintaining only ten Negroes, was $15,285 for 48,011 days, and in the following year $19,033 for 62,872 days. The increased labor was needed to harvest and process a bumper crop that yielded nearly 300 tons of sugar, with most of the increase accounted for by 9,000 additional days worked by cane-cutters and trapicheros. The average labor costs for the two years were about $17,000, as they had been in the three years 1718–21, suggesting that the scale of labor inputs had not changed much in more than 40 years.

In terms of labor inputs, the 7 years of full data and unexceptional circumstances between 1780 and 1822 represent a return to the high labor inputs of the decades before 1600 and the abolishment of the repartimiento. The average total number of days worked annually in the earlier period was 79,795 and in the later 80,682. Major differences existed, however, in the average absolute cost of labor, 50 per cent greater in the later period, and in the average annual labor bill, which had risen by approximately $8,000.

Relation of Labor Costs to All Costs. In order to calculate the percentage of all costs accounted for by payments to workers, rations, and the costs of Negro slaves, I have used the data of accounts of four periods in the accompanying tabulation. These show that nearly two-thirds to three-quarters of all outlays for the mill went to supply labor for it; the estimates of total costs are probably accurate enough to enable us to conclude that

	Labor Costs/All Costs
1591–1600 (9 yrs.)	75%
1622–1624 (3 yrs.)	70
1768–1795 (7 yrs.)	67
1811–1831 (5 yrs.)	62

the costs of labor accounted for more than half of all costs through its history.

Productivity of Labor

The simple definition of productivity that I have used is the pounds of sugar made per unit of time, here the workday and year. The lack of data presents some problems. There are many years of data either on total production of sugar or on labor, but not both; for only 23 years are both available. The volume of molasses produced was never described, yet it was a major product of mills like Atlacomulco. In some years assumptions have had to be made concerning the rate of pay of Indian laborers. However, it seems useful to know at least the order of magnitude of productivity, so I have summarized the available data and present the results in Table 32 (p. 135).

Some of the assumptions made to calculate these figures should be discussed here. In the case of Negro slave labor, I have assumed a work year of 300 days and multiplied this figure by the number of Negroes over the age of 10 and under 60. The contribution of young slaves may have been balanced by the time that Negro mothers spent caring for younger children; in addition, it appears likely that older persons contributed less than persons between the ages of approximately 25 and 45, as the data on changes in value associated with changes in age suggest. The total days worked by Negroes may thus have been overestimated, operating to depress estimates of productivity in the years when they were important in the labor force, as well as to understate the daily cost of maintaining Negro slaves.

Calculating the contribution of free labor is somewhat easier, partly because no women and very few boys were employed except in the purgery. If the daily rates of pay received by free workers are stated, the total number of workdays may be calculated easily from accounts in which are distinguished the amounts of money paid to Indians and other day laborers. The smallest total amount of workdays is attributable to supervisory personnel, and I have assumed that many of these — mayordomos and field supervisors — worked 350 days annually rather than 300. In any case, after about 1605 the number of supervisors was very low.

Four periods may be distinguished on the basis of the results. In the first, based on the most nearly complete series of records from 1581 to 1602, production did not rise above 2 pounds of sugar per man and day, fell in one year below 1 pound, and averaged only 1.4 pounds

of sugar per man and day. This result stands in marked contrast to the average production of 2.1 pounds per man and day for the years of low total production and labor inputs from 1620 to 1624, when productivity rose above 2 pounds in three of the years, and never fell below 1 pound. Productivity in the two years 1644 and 1645 did not reach the high average of the previous short period, amounting only to 1.9 and 1.6 pounds; the higher of these values, however, was not equaled in the period 1583–1600 and the lower was exceeded slightly in only 2 of the 9 years in that period.

In the 8 years from 1768 to 1811 for which sufficient data are available to calculate, productivity reached values never previously attained: whereas the highest value of all the previous years was 2.8 (1621), the lowest value of these later years was 3.4 (1779) and the highest 9.3 (1769). On the annual scale differences of magnitude exist that dramatize the changes. For the four short periods discussed they are, respectively, 411, 633, 513, and 1,764 pounds per man and year. No problems associated with calculations of inputs of Negro labor, which may be overestimates, can account for the differences between the first three and the last figures; furthermore, the technique, discussed below, used by the administrator Urquijo to disguise a rise in the cost of labor in the late eighteenth century introduces a bias of opposite sign, with the result that the figure of 1,764 may be an underestimate.

The details of labor inputs of the late eighteenth century have great interest because they show not only differences in inputs of different kinds of labor, but also differences in combinations of labor employed as time passed. These changes seem to be trends; so I have placed them on a graph (Fig. 19) to see if the trends can be related to increases in production. Naturally there are difficulties. In the first place, there are too few years with the proper data, and in the case of sugar cane the importance of this fact lies in the dependence of sugar production in any year on labor inputs of the previous year. In the second place, the effect of the Independence movement and the severe and frequent frosts of the second decade of the nineteenth century resulted in a diminution of the expected correlation between labor inputs and production of sugar; nevertheless, although uncertainty is the major feature of agricultural operations, the variability associated with it must here be regarded as a separate problem. Finally, there is the difficulty introduced by Urquijo's practice, referred to above, of increasing the stated number of

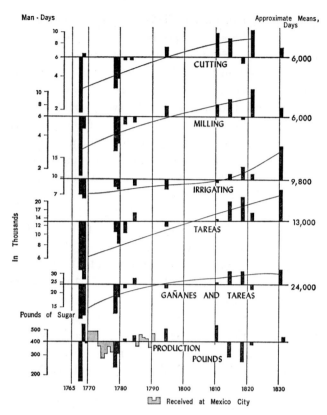

Figure 19. Inputs of various kinds of fieldwork, 1768–1831

workers above their actual numbers in order to account for larger total wage payments rather than to state the true wage rates and true number of days worked. Because of these deficiencies in the data, simple graphical analysis rather than more elaborate techniques seems most suitable.

The graph shows that from about 1765 to 1795 production of sugar and deliveries of sugar to Mexico City fluctuated about the figure of 200 tons, approximately the mean of production for all the years from 1768 to 1832. The less abundant data concerning five important categories of labor inputs suggest, on the other hand, that labor inputs were increasing over the entire period 1768–1831, so in fact the intention must have been to increase total output.

Ten categories of labor account for approximately 90 per cent of all labor inputs for the 12 years of available data from 1768 to 1831, and of these the six most important are shown in Figure 19 with their approximate mean annual inputs in man-days. All exhibit similar variation throughout the period, except that irrigating exhibits less relative change than the others. Cutting and milling, as might be expected, are closely correlated; they are, besides, similar in magnitude. It is surprising that transport does not vary directly with them. Tareas and gañanes pose problems of description, in that it is not clear how they may be differentiated; for this reason I have both combined them and shown tareas separately. Their long-term tendency was opposite, with gañanes considerably larger than tareas in the beginning (for example, in 1769, 8,793 gañanes but only 3,907 tareas) but considerably smaller by 1831 (9,623 gañanes, but 24,890 tareas). Thus the change in the importance of tareas is the greatest change shown on the graph, but when combined with gañanes, which had declined in importance from a peak of 13,285 in 1815, the change in relative importance of the group was not so great. Both of these categories involved plowing, cultivation, and the elaborate irrigation procedures. Thus, a large part of the input of these two categories contributed to production in the following rather than in the current year.

Evidently, questions concerning increases in efficiency in various phases of plantation operations cannot be settled without data for successive years, since so much of the labor input in a given year is related to output in the following. Yet it would be extremely interesting to know the magnitude of variation in output in an agricultural enterprise that ignored fertilizers, had no advantage through varietal change, and little control over variations in such environmental features as insect pests, rodents, and plant diseases — in other words, whose major input and independent variable was labor. Perhaps the graph shows that the intention of the owners and managers was to increase output by increasing labor inputs, but that circumstances beyond their control operated to prevent realization of their goal.

X

Summary and Conclusions

WHEN Hernán Cortés decided to establish a sugar plantation near Cuernavaca, he added to his many other properties and rights an enterprise that subsequently became the major agricultural property of his heirs in Mexico. In the first century of its existence it prospered from the high and rising price of sugar, but, like many other economic ventures in New Spain, it fell upon hard times that began in the seventeenth century and persisted through much of the eighteenth. Although there was recurrent discussion concerning abandonment or sale of the plantation, it nevertheless remained part of the Estate and finally experienced a second, but brief, period of great prosperity during the California gold rush (see Appendix H, p. 121).

The Estate of which it constituted a major part, and which carried it through bad times, derived most of its income from tributes and from urban properties in Mexico City. It was a complex institution of relatively great size in the sixteenth century, but after seventeenth-century alliances of the descendents and heirs of Cortés with the Neapolitan Dukes of Terranova and Monteleone the properties of the Estado del Valle came to be overshadowed in value by the European holdings of the heirs; their properties in Spain alone, for example, were sold in the nineteenth century for more than was received shortly after for the Mexican rights and properties, excepting some urban property and San Antonio de Atlacomulco.

Cortés found himself in the 1530's limited in choice of sites for his mill and canefields by the still dense population that existed in the region, and by the prior claim to water for power and irrigation established by Antonio Serrano de Cardona. The site occupied by Cortés' mill for a century was thus inferior in numerous respects, the soils used to grow cane were not the best in Morelos, nor was the distribution of the canefields optimum, scattered as they were over a very large area. But time and circumstance were on the side of his heirs, and as the population declined opportunities arose to occupy a better site and to use fields and pasture that formed a single unit. Although rental of land had been important in the sixteenth century, by the middle of the following century this mode of tenure declined in importance, and by the mid-eighteenth century the shape of the cultivated area not only approximated the form visible on the map of 1850 but also enclosed land mostly owned by the Estate. In addition, the main outlines of the water conveyance and distribution system, laid out in the sixteenth and seventeenth centuries, were completed in the mid-eighteenth century. The problems concerning land used for pasture and milpa that arose in the nineteenth century had their origins in part in population increase, in part also in intrigues fomented by individuals who, using the name of Indian communities, wished to control the community land for their own purposes, and in a general unrest that seemed to increase after Independence; these conflicts were never fully resolved in the nineteenth century.

The Cortés plantation has been analyzed in this work in respect of its equipment and stock, labor, organization, and yields over a relatively long span of time. Perhaps the most striking of the observable changes is the increase in the amount of sugar produced per unit of labor, which rose in the early seventeenth century over the amount produced in the late sixteenth century and then by 1768, when records again become adequate for analysis, increased yet again and markedly over earlier amounts. Yields per acre of sugar also increased over time—in the absence of varietal change, this was probably due to a greater extraction rate rather than to bulk of cane produced.

The change in amount of sugar produced per unit of labor was approximately fourfold in the period reviewed here. In an agriculture characterized by widespread application of scientific research, a change of such magnitude would be regarded as the near-equivalent of a revolution, a sign of progress and great improvement. Yet it seems unlikely that even those involved in the industry knew that it had happened; this is the implication of Ruiz de Velasco's characterization of the mode of production that held sway in the late nineteenth century in Morelos as Cortesian, and there is no sign in his work that he saw any difference between the sixteenth and nineteenth centuries in respect of productivity and efficiency.

Probably numerous sources contributed to these increases. On the one hand, there occurred a group of unpredictable shocks or disturbances, in relation to which the improvements may be interpreted as response: to this class belong the fall in the price of sugar that began about 1600, after a long period of rising prices; the abolishment of repartimiento labor for sugar haciendas, occurring at about the same time; and the related depopulation of New Spain, apparently made conspicuous by the results of the great epidemic of 1576. These circumstances, coupled with the inability of the Negro slave population to reproduce itself—at least not on this plantation—must have suggested the alternatives of reorganization or abandonment of the enterprise. In the short run, the response was both to curtail production and to increase slightly the productivity of each input of labor. Probably the latter could have been achieved as well in the 1580's as in the first two decades of the following century, but the need to do so did not exist until the repartimiento was abolished. Over the long run, a different kind of labor, and different labor relations, developed that enabled the Cortés plantation to remain in operation: the free, resident, mulatto or mestizo laborer, who lived on the plantation, received a weekly or monthly wage, and raised most of his own food on plantation land, supplemented the Negro slaves and Indian workers who previously dominated the labor supply.

This class of labor seems not to have been important in the mid-seventeenth century, as shown by the dis-

cussion in the Valles case concerning compensation for missing Negro slaves: there was no mention at that time that such a group provided an important alternative either to Negro or part-Negro slaves, on the one hand, or to Indians on the other. The attrition of the mulatto slave ranks at Atlacomulco by flight shows where at least some of the free population of Morelos originated; between the late eighteenth and nineteenth centuries the size of the free resident population doubled. A sufficient index of their differences from Indians, for purposes of this study, is given both by their preference for payment by the task rather than the day and by the preference for the faster mule rather than the ox in plowing. Some part of the change in efficiency must be attributed to the emergence and growth in size of this group of people.

Another source of increase in productivity, which is more difficult to establish but which nevertheless seems to have been operating, was a gradual increase in the level of skills and insight. This was expressed in improvement in the level of performance of routine tasks, whose cumulative effect over one or two centuries was probably very important. Here we find several lines of evidence. In the first place, the accounts of the plantation show great improvement in respect of organization, if not purpose: by the mid-eighteenth century, their form clearly expressed recognition of the overwhelming importance of labor costs, which in the sixteenth century appeared in the accounts simply as a few items of cost among many, some of them nearly completely trivial. In the second place, there occurred a rapid substitution after 1600 of non-Spanish labor for Spaniards in positions that apparently had inevitably to be staffed by Spaniards in the sixteenth century. To be sure, in the first century of operations Indians and Negroes had been employed as maestros de azúcar, but at least one Spaniard had always supervised them. After about 1600 the amount of supervision was reduced, and the qualifications necessary for various tasks were acknowledged to be more widely held than had previously been the case. Another line of evidence has to do with the improvement in the mayordomos, shown most clearly and simply in the letters they wrote to Estate officials. Their letter-writing ability increased over time, and by the early nineteenth century several of them had ample command of the complicated and elegant mode of letter-writing that had become fashionable in the eighteenth century.

Another group of changes occurred in fieldwork and

in equipment at various times in the eighteenth century and should be regarded as related to changes in productivity. In the fields, these changes occurred: the planting dates became stabilized in the latter part of the eighteenth century; only plant cane came to be used in milling; cane was harvested at a younger age than it had been before; and a greatly increased acreage was harvested. Also important as indicators of change in field routine were the use of the standard — but subsequently frequently enlarged — tarea, and the use of permanent names for fields beginning about 1750. The numbers of coas and rejas castellanos on hand at inventories increased greatly over the numbers previously held. A result of these and other changes was that heavier yields of sugar were obtained than ever before.

Within the mill itself, the use of bagasse presses ceased sometime in the 1750's, probably pointing to an improvement in the amount of juice extracted by the mill, but little information is available on changes in the latter. The average weight of kettles doubled between the inventories of 1721 and 1746, and the number of kettles increased by two to a total of ten. In the 1740's, the mill was converted to water power after a brief period of use of mule-drive.

Considered individually, none of the above changes in equipment and routine would be regarded as very significant; considered together, and in relation to the greatly increased productivity of labor and yields per acre, it appears that they were important. Yet none of them stands singly as an important technological change, not even the doubling of the average weight of kettles. Each is instead a relatively minor change, affecting only a single part of a complex system, and none represents an addition to the system in the form of equipment or its use.

The data and conclusions raise questions concerning the rates and natures of the changes that occurred, but these are difficult to answer with the Atlacomulco data, because the records are discontinuous. The effect of the discontinuity is to make the results for the years after about 1750 appear the result of sudden rather than gradual improvement, hence suggestive of revolution. Here data for the years from 1680 to 1740 are required but unfortunately not available. However, the problem may be approached in a different way, by considering the very great number of elements that contributed to the changes in productivity: Figure 19 (p. 102), showing variable relationships among different kinds of labor in the late eighteenth and early nineteenth centuries, is a useful reminder of the complex nature of the relationships. It seems likely that the numerous changes in operations and equipment occurred semi-independently of one another — there is no obvious connection, for example, between the sizes of kettles and the standard muleload — and this characteristic should be sufficient cause to expect that some part of the changes in productivity were the cumulative result of small improvements in unrelated aspects of the operation.

The problem of diffusion of technical information within New Spain and the entry of knowledge there from other places must also be dealt with indirectly. Enough has been presented to show that the repertory of information and skills applied to sugarmaking at the Cortés mill was not significantly different from that applied at any other mill in Morelos, if only because of the way supervisory personnel and lessees moved from mill to mill. In addition, it was shown in other connections that this mill was a focal point of importance in the region: not only did laborers come and go, but nearly everything used at the mill was purchased elsewhere and carted or carried to it, and weekly demands for cash required the maintenance of links with merchants in Cuernavaca — it was, in short, part of a complex and extensive network made up of acquaintanceship, obligations, and traditional relationships through which flowed both goods and information, oral and written. The access to information ensured that the management of the mill followed the pattern of the region.

The only significant technical difference between Morelos and West Indian mills, apart from the use of the bagasse press in the former, was the refusal to adopt the Jamaica train in Morelos. Since the Jamaica train was used in the West Indies after about 1700 largely to conserve fuel, and most persons felt that its use reduced control over the boiling process, even this difference did not necessarily place the Morelos producer in a disadvantaged position. In respect to the world sugar industry, the Morelos segment has to be regarded technologically as part of the whole: it was started by and depended upon persons familiar with the relatively well-developed industries of the Canaries and Santo Domingo, similar to the Brazilian industry that provided the common technical base on which the West Indian industry was built in the middle of the seventeenth century. The fact that thereafter little technical change occurred until after 1800 makes all the more interesting the fact that, at least in Morelos, significant changes in productivity and yields did occur.

APPENDIXES

Appendixes A–H

A. POPULATION CHANGE IN MORELOS

FOR the time before 1795, the only available indicators of population change are the changes in the numbers of persons paying tribute, listed in official counts. These lists cover most of Morelos except Cuautla, which may have been under the jurisdiction of the Estate only a short time. I have seen no original reports or autos concerning the official tasaciones, but have relied instead on statements in various annual accounts of the Estate that describe the date of approval of the count and the number of tributaries counted. None of these statements defines a tributary, or includes any notice concerning changes in the definition of a tributary, if such changes occurred. Furthermore, none is complete in the sense of describing individually all the pueblos tributary to the Estate; for example, in most cases the sixteen or so pueblos described as sujetos of Cuernavaca are simply grouped under the headings "Cuernavaca" or "Villa de Cuernavaca." The only exception I have seen is the more or less complete listing of pueblos dating from 1568 and published by Borah and Cook (1963). In the late seventeenth and early eighteenth centuries there appeared in the accounts numerous complete listings of the fourteen pueblos of the Tlalnahua or Tlalnahuac, which, although only rather small sujetos of the Villa de Acapistla, were nevertheless treated in great detail. In fact, these fourteen pueblos are treated as a unit from the time that reasonably full records began to appear in the late sixteenth century, and their distinct identification suggests that they must have had a very interesting history. All fourteen still exist, except Tetehuama, described as "despoblado" in L238 E4, and its abandonment probably occurred shortly before 1600. The remaining thirteen, their names spelled as on Tamayo's map (*Carta General del Estado de Morelos*, no date, but probably 1959 or 1960) are: Jantetelco, Amayuca, Jaloxtoc, Tlayecac, Tepalcingo, Anenecuilco, Atotonilco, Axochiapan, Tetelillas, Telixtac, Chalcatzingo, Jonacatepec, and Atlacahualoya. The pueblos of the

Tlalnahua occupy the warm, dry plains of southeastern Morelos.

The following list of sources of data for the graph of population change (Fig. 2) does not include a complete listing, by pueblos, of all the units tributary to the Estate in Morelos except for 1791 and 1885:

Year	Source
1568	Borah & Cook, 1963
1583–94	L260 E10
1595	Borah & Cook, 1963
1615	L237 E14
1629	L237 E10
1635	L237 E14
1638	L237 E12 or E17
1606–55	L238 E4
1680	L238 E17
1686	L28 V49 E12
1701, 1707, 1715, 1721, 1726, 1732	L179 *Pergamino*
1721	L28 V50 E6
1769	L298 E81
1785	L298 E14
1791	Mazari
1806	L244 E17
1885	Velasco

There is uncertainty concerning the date of the "Padrón Itinerario" published by Mazari in 1927, of which the original is in the Bibliothèque Nationale, Paris. Although he attributed the list, complete for Morelos except for the jurisdiction of Cuautla Amilpas, to the year 1695, the attribution is incorrect. The names of persons described as owners of various haciendas all date from the late eighteenth and early nineteenth centuries, of whom the best known are Don Gabriel Yermo, owner of San Gabriel and numerous other properties in Morelos, and Don Nicolás de Icazbalceta, owner of Santa Clara. Because of the inclusion of these and other personal names, it should be relatively easy to fix more exactly the date of this census. Furthermore, Volume 8 of Padrones contains a map of Morelos and a Padrón itinerario of Cuautla very similar to the one published by Mazari, except that the Cuautla Padrón contains more information on the sizes of the various castes. The map contains all of the places, even very small ones, listed by Mazari, and it appears unlikely that it is anything but the map made for the Padrón. Padrones V98 E2 contains a letter from Domingo de Vitorica to the Intendente of the Province responding to the latter's call for a census; Vitorica's letter is dated 11 March 1790, and Vitorica stated that the count would begin as soon as the clergy (vicarios) had been instructed, a point requiring considerable attention because they were, as he put it, so ignorant. The census had already begun in some places, since he also reported that some persons fled their houses owing to their misunderstanding part of the questionnaire. Because the details of his letter seem to refer to both the map and the list published by Mazari, I conclude that both date from 1791 or 1792.

Velasco published a complete list of the population of Morelos by municipal districts. Although the date of the census was not given, it probably dates from about 1885. The 1950 figure is contained in the government census of that year. Estimates of the 1959 population given by Tamayo suggest that there was a rapid increase in the rate of change in the period 1950–59.

B. LESSEES OF THE PLANTATION

THE plantation was administered by the Estate from its founding in 1535 or so until 1 January 1567, when it was leased to Miguel Rodríguez de Acevedo for 9 years and managed in his name until 31 December 1575. The contract between Rodríguez de Acevedo and the Marqués (L107 E3) does not contain a list of his bondsmen, so it is impossible to state his connections or evaluate his importance in New Spain in those terms. However, he was a vecino of Mexico City (*ibid.*) and, more important, founded with his wife Catalina Pellicer y Aberruza a mayorazgo in 1574 (Fernández de Recas, pp. 199–201) that comprised a house in Mexico City in the Calle del Reloj, the hacienda San Miguel del Bosque and molino del Matadero, both in Atlixco, Puebla, and another mill called La Candelaria. The information given by Fernández de Reca suggests that their descendents had become related to the holders of the titles of Marqués de Santa Marta and Conde de Quintanilla by the seventeenth century. Miguel Rodríguez de Acevedo himself had had business dealings with the Estate at least as early as 1565 (L267 E1 D19).

For some reason, Estate officials did not wish to resume administration after his tenure, and looked instead for someone willing to lease the plantation for five years. Mateo Rodríguez agreed to do so, and his lease began on 1 January 1576 and ended 31 December 1580. The list of his bondsmen is included in the document containing his contract with the Estate (L282 E2). Among those who posted 2,000 ducats each was the well-known Gordián Casasano, himself owner of a sugar plantation near Cuautla Amilpas that was one of the first established in New Spain, as well as occupant of important government positions and an official of the Estate. Another was Diego López de Montalbán, renter of Tuxtla and some estancias from the Estate until at least 1576 (L228 E3); he had also witnessed the signing of the contract between Rodríguez and the Marqués in 1575, as had another of the bondsmen, Juan de Moya. The latter had been co-bidder with Rodríguez in the latter's initial bid for the plantation on 1 December 1575. Others were Pedro de la Palma, Pedro Pérez, Pedro Rodríguez Patas, Hernán Bázquez de Aldana, all vecinos of Mexico City; Baltasar Bañegas, a vecino of Cuernavaca, and Pedro Gutiérrez and Miguel de Haro, whose residences were not given (L282 E2). The involvement of Mateo Rodríguez in the sugar business is clear from other information: for example, in the 1586 account of Tlaltenango (L262 E2 D6) his name appeared in a list of names of persons whose cane had been milled on shares at the

mill, and there is a notice in L208 E7, dealing with conditions of 1586, that a nephew had a sugar store in the Calle de Tacuba in Mexico City, where he received sugar from his uncle. Nor had Miguel Rodríguez de Acevedo, his predecessor, operated alone; evidence of this is the fact that in 1573 (L218 E3) Alvaro Rodríguez de Acevedo bought 1,625 fanegas of maize in Yautepec for Tlaltenango, suggesting that he was managing the plantation.

The renter Andrés Arias Tenorio, whose lease extended from 1625 to 1634, was deeply involved in the sugar industry, as were at least some of his sons. The first mention I have found of him occurs in the account of 1613, when some of the cane from the Tlalcomulco fields was processed at his mill, Amanalco, rather than at the more distant Tlaltenango. He performed the same service in 1614, and Tlaltenango reciprocated in 1616–7. Thus at least by 1613 he was the owner of Amanalco, a small mill started by Bernardo del Castillo, perhaps in the 1530's; del Castillo began with a very small piece of land, 300 by 200 varas, added to in 1536 so that its area was then perhaps doubled. I have no record of the acquisition of this land by Arias Tenorio, but in 1618 he was granted 4 caballerías by Pedro Cortés, and in 1619 12 caballerías more (the total was more than 1,600 acres); in 1616 he acquired from Pedro Cortés 1½ sitios de ganado menor, so that by the time he rented Tlaltenango he was well known to the officials of the Estate. Later, a relative, Francisco Arias Tenorio, vecino of Cuernavaca, increased these holdings by obtaining in 1640 one caballería from Alvaro de Acevedo, heir of Miguel Rodríguez de Acevedo (L28 V50 E2). Much or all of this property was finally grouped into three units: the two sugar plantations, Santa Ana Amanalco and San Francisco Pantitlán, and the ranch called Michiapa or Michapa, all between Cuernavaca and Cuautla. These properties seem to have been lost, perhaps in the 1640's or 1650's, through debt to the Tesorero Antonio Millán, at one time "General de la Santa Cruzada [of New Spain], sus provincias, y Islas Filipinas" (T1272). At the time the rental was agreed upon between Arias Tenorio and the Marqués, however, the business affairs of the former were prospering, and no difficulties of any moment were encountered when the lease terminated in 1634.

Captain Carlos García de Covarrubias leased the plantation following the rental of Arias Tenorio and, as had Tenorio, died shortly before the rental period ended on 20 April 1643. In 1612 he was a resident of Sevilla (L325 E6), where at about the same time he married Doña Ana Carillo; the date of his arrival in the New World is not stated in this document. However, one of his bondsmen was Hipólito de Santoyo, who was also controller of the Estate and who took over the administration of the mill when Covarrubias died. The major feature of their rental period was a lawsuit at the end for the amount due on the last nine months ($3,666-5-4), but whether or not the amount was collected I do not know. It was also during their rental that the mill was moved from Tlaltenango to the site now occupied by the ruins at Atlacomulco.

After a brief period of administration by the Estate, the plantation was leased in 1647 to Antonio de Araña and his cousin Pedro de Egurén, the latter a vecino of Mexico City and Cónsul más antiguo de la Universidad de los Mercaderes; possibly in this case Araña ran the plantation, and Egurén sold the sugar. Only the rental contract between the two cousins and the Estate and the inventory with a short lawsuit exist, so that other information concerning their activities is not available. It appears that Araña had bought slaves, oxen, carts, and other equipment from another plantation near Cuernavaca, and wished to put them to use at Atlacomulco (L295 E137).

Events are obscure for the twenty years following the end of the Araña-Egurén rental, from 31 December 1654 to the beginning of the rental of the Alférez Onofre de Vargas on 1 January 1673. The history of the de Vargas rental illustrates well the interdependency of the history of the mill and the Estate, and the story is rather complicated. The Estate had been rented for the ten-year period 1671–80 to Captain Don Francisco de Avello, who died in 1676; his widow, Doña Béatriz de Tello Sandoval, obtained permission to continue with the lease after his death, but went bankrupt in the same year, announcing the event by taking refuge in the Cathedral in Mexico City. In the same year Joseph Ruiz and Luis de Herrera, the latter a maestro de carrocería in Mexico City, where he owned some property, sublet Atlacomulco from Onofre de Vargas (L93 E14); as far as the Estate was concerned, they were co-renters with de Vargas (L93 E7), since the Estate had nothing to do with the new contract, yet de Vargas himself played little part in the subsequent litigation. Perhaps, since in 1672 he was acting for the Governor Avello in Mexico City (L93 E3), he had merely been running the mill for the governor. Béatriz de Tello Sandoval's bankruptcy in 1676 apparently did not affect Ruiz and Herrera, for they arranged a new agreement with Pablo de Fajardo Aguilar, who had himself newly assumed the Avello-Sandoval contract with the Estate, to run until 1680. Fajardo died apparently shortly before the end of his lease, and on 30 May 1679 Joseph Ruiz, hearing that he was about to be arrested for failure to pay part of the rent due on Atlacomulco, took refuge in the church at Yautepec, whence after a few days he fled to the Cathedral in Mexico City, where he was joined by Luis de Herrera. Some of the possessions of the latter — Ruiz appeared to own little property — were sold at auction in 1680 to satisfy a debt of approximately $12,000 owing to the Estate: $7,171 for deficiencies of cane, and $5,049 for other items missing at the plantation, including some slaves. At least some of the money was regained by the Estate, and the plantation then continued in administration until 1683, for part of the time in depósito — that is, accountable to the courts while the litigation proceeded — under an administrator or depositary, Francisco de Salvatierra. It is a pity that more information is not available concerning the relations among de Vargas, Ruiz and Herrera, the Avellos, and Pablo de Fajardo, since the few details sketched above suggest an interesting set of linkages that may be typical of late seventeenth-century business ventures in New Spain. Particularly interesting is the case of Luis de Herrera, a tradesman, yet the owner of a substantial amount of property, father-in-law of Joseph Ruiz (L93 E2), and one of the bondsmen of Béatriz de Tello Sandoval (L93 E7). It seems to have been in large part his financial difficulties, involving a shortage in the accounts of the estate of one of his fellow carroceros (L93 E2), the subject of litigation for some time, that touched off the failure of his partnership with Ruiz.

The mill remained in administration or in depósito from 1 September 1680 until 1 January 1683, when a new renter, Captain Juan García del Castillo, a vecino of Cuernavaca, entered into a nine-year lease to run until the end of 1691 (L30 V53 E1). García del Castillo was familiar with the sugar business and with Atlacomulco, since he had been the renter (L30 V53 E2) of Amanalco in 1680, or its administrator (L93 E14), and had been appointed (L93 E15) to represent Ruiz and Herrera at the inventory of Atlacomulco of 14 June 1679, following hard on the heels of their bankruptcy; in addition, in 1694 (L93 E21) he was renting the ranch Mazatepec from the Estate, on which he owed $1,000, and was either owner or renter of the adjacent ingenio San Salvador Miacatlán. Oddly, he appointed Joseph Ruiz to act for him in Mexico City in connection with the negotiations for the rental of Atlacomulco (L30 V53 E1). García Castillo gave up the rental in early November 1688, and the Alférez Esteban Súarez Narciso assumed the lease for the remaining two years and two months, as well as additional time for a total of nine years. The latter, a vecino of Tlayacapa in Morelos but resident of Chalco (L93 E12), had attempted unsuccessfully to rent Atlacomulco in 1680 after the failure of Ruiz and Herrera, as also had García del Castillo; probably it was the litigation that forced the plantation into depósito that prevented one of these bids from being accepted, as well as questions by Estate officials concerning the amount of bond necessary.

The bondsmen of García del Castillo included Juan Ruiz de Colina, who bore the family name of a former governor of the Estate, Captain Buenaventura de Barrientos, and Joseph Núñez de Contreras, the latter at the time renter of the mill Santiago Zacatepec, outside the town of the same name in southern Morelos; all were said to be well known and of good credit (L30 V53 E1). García del Castillo agreed to be a bondsman of Súarez Narciso, to the amount of $2,000. Other bondsmen were Bartólome Núñez de Contreras, renter of a trapiche de hacer panocha in Yautepec called Apisaloque or Apizaco; Captain Luis Domínguez, renter of the ingenio Temilpa near Zacatepec, owner of the ranch Amatitlán, outside Yautepec, and in 1677 representative of the governor Fajardo at the inventory of Atlacomulco (L93 E15); and Andrés de Nicolae (L93 E8). Súarez Narciso paid the rent through 1690, but in 1691 paid only part, and in 1692 nothing (L148 C3), so that early in 1692 the Estate sued for $6,396-6-0 owing. He offered as payment only 7 slaves, 224 horses, and 550 loaves of sugar, an insufficient amount, with the result that he was put in jail as a reo executado, and Joseph de Bárcena, a vecino of Cuernavaca, was made depositary of the possessions of Súarez Narciso and of the mill. On arrival at his ranch outside Yautepec the officials discovered that Domínguez had fled with his family and his few possessions a few days before; he was discovered five days later in refuge in the church at Jonacatepec. By November things had worsened considerably for Súarez Narciso, since more rent was owing, he had been in jail some time, and he could not afford to hire a mayordomo to run the mill, a post that he had previously filled himself.

As the Estate proceeded against the bondsmen, their shaky financial positions were revealed. In early 1693 García del Castillo, then renter of Mazatepec, informed officials that all his possessions had been seized by the Inquisition, except six Negro slaves, whom he was ordered to bring to Cuernavaca. Bartólome Núñez de Contreras took refuge in the church at Yautepec (L93 E8), after which it was discovered that all he owned was 1,500 varas of cane planted at the mill Apizaco that he had been renting; he was persuaded from the church and taken to the Mexico City jail by Don Cristóbal de Bárcena, vecino of Cuernavaca and later to assume the post of mayordomo at Atlacomulco. The Estate then moved against Andrés de Nicolae, vecino of Mexico City and possessor of a letter from the Viceroy stating that since he was

111

a noble (hijodalgo viscayno) his possessions were not liable to confiscation (L93 E8); in any case, he was subsequently jailed for debts to the Crown, at which time he was revealed to be penniless. In Carrión, Atlixco, where Luis Domínguez had lived, officials found that he had died in June 1693, leaving his family in "tan suma pobresa" that they were living on alms from the Bishop of Puebla and begging for money to build his tomb; the case against Domínguez was dropped. Meanwhile, the jail in Cuernavaca where the confiscated slaves of García del Castillo had been held burned to the ground and the slaves escaped, taking to the road as highwaymen; although the Governor of the Estate caused "censuras asta la de Anáthema" to be read against them in the church at Cuernavaca, they had not been found by May of 1695, and the case seems to have ended there, with $1,000 paid by Núñez de Contreras, some cane, slaves, and horses obtained from some of the other principals, the ruin of nearly all the bondsmen (L93 E8), and probably $10,000 lost by the Estate. There seemed little more to do, so Governor Morales was freed from further responsibility. Súarez Narciso died of natural causes in jail in Mexico City on 4 May 1694, owing approximately $25,000 to the Estate, according to the calculations of the controller, as well as debts to others.

The mill was declared available for rental in late 1692, and in the following February the governor announced that it had been rented to Francisco García Cano, vecino of Mexico City, and Diego Urtado Sagache, vecino of Cuautla Amilpas; the rent was low — only $3,000 and four cargas each of sugar and molasses per year — because of the condition to which it had sunk as a result of the difficulties of Súarez Narciso and a recent frost that had damaged much of the cane. García Cano and Urtado assumed control of the mill in 1693, the lease to run for the customary nine years, or until early in 1702. Apparently there were no difficulties associated with the rental payments or the final inventory, since Diego Urtado continued as mayordomo when the Estate resumed administration of the plantation at the end of their contract. Urtado died suddenly in 1706 (L30 V56 E1) with a small shortage in his sugar and molasses account, and the administration was accepted by Cristóbal Mateos until 1709, when he rented the mill.

García Cano's ties with the plantation community were strong. He was married to Antonia de Aranda y Echavarría (T1979 E9), a member of the Aranda family that owned the trapiche Nuestra Señora de Guadalupe in the first part of the eighteenth century and Temixco in the seventeenth and eighteenth centuries (L447 E10), as well as Amanalco (L44 V79 E12). Diego Urtado had apparently previously acted for García Cano in the purchase of a small trapiche, San Nicolás Guaumecatitlán near Zacatepec about 1700, or else had owned it before García Cano took it over on his death in 1706; to expand this trapiche García Cano obtained a grant of three caballerías with water for irrigation in June 1707 (L447 E92).

The next renter, Cristóbal de Mateos, died sometime before the end of his lease in August 1718, but his widow, Doña Juana del Villar y Arena, decided to continue until the end of the contract. Mateos had been owner of the small mill Atotonilco in southeastern Morelos. When in 1718 his widow turned the plantation back to the Estate, which then administered it through Cristóbal de Bárcena, there were claimed to be some shortages, but the matter was soon settled in her favor.

When the mill was offered for rental in 1719 one of the bidders was Don Manuel Osorio, vecino of Cuernavaca but resident of Mexico City, who offered $3,500 for nine years. For some reason the Estate officials did not act on his bid. In August 1721 Joseph de Palacios, vecino and mercador of Mexico City, bid $2,000 annually for a tenure of six years, offering as bondsmen six other vecinos and mercadores of Mexico City. De Palacios was a person of "notorio caudal," owner or lessee of the mills called El Puente, Temixco, and Miacatlán (T1965) and Alcalde Mayor of Cuernavaca (L345 E3). His bid was accepted.

At the inventory de Palacios was represented by Manuel Osorio, listed in this document (T1965) as a vecino of San Vincente, a mill a few kilometers south of Atlacomulco. There were no problems associated with the rental of de Palacios, who continued in the lease for ten years, until August of 1731, and left improvements valued at $1,385.

The small credits claimed by Doña Juana del Villar and de Palacios were ominous portents, since the claims of the two renters that followed the latter were much larger and resulted in extensive litigation. Put simply, the rental price had fallen so low that most of it was used for maintenance and repairs, leaving no hope of profit.

The next renter was Antonio Delgado Matajudios, vecino of Cholula, who entered into a nine-year lease extending from 1731 to 1740. The facts surrounding both his rental and his abandonment of it are obscure, but at some time in the nine years the mill was taken over by one of his bondsmen, Joseph González de Movellán y Lamadrid, Régidor Decano y Contador de Menores y Albazeazgos of Mexico City. De Movellán y Lamadrid was also a bondsman of Augustín del Linal, who had contracted to supply meat to Cuernavaca, and he had assumed Linal's contract as well. He gave up the mill in 1741, and early in 1742 the controller of the Estate calculated that he owed approximately $12,000, of which $10,000 was due on the plantation and $2,000 on the meat supply. The inventory of August 1741 showed that most of the buildings were in very poor repair and the roof of the milling house had collapsed (T1965). By 1752 the case had been resolved and an Estate official wrote the Juez Privativo that de Movellán was not a debtor to the Estate (L93 E11); in fact the Estate may have incurred a loss of more than $10,000 on this rental (L341 E14).

Tomás de Avila Romero leased the plantation in 1741, at which time he may have had a store where sugar was sold in Mexico City; certainly at some time in the 1740's he was proprietor of such a business (L144 E8). His first difficulties with the Estate concerned compensation for a new aqueduct and water intake that he built, and as a result of the litigation which ensued the mill was temporarily placed in charge of Pedro Bermúdez de Orense, who claimed to have given de Avila Romero $6,000 to run the mill, taken from the Cuernavaca tributes by order of the Governor Bermúdez de Sotomayor; de Avila Romero apparently was absolved of responsibility in the matter after it was discovered that the money was part of a much larger sum, about $60,000, that Bermúdez de Sotomayor managed to transfer to Bermúdez de Orense without adequate records concerning the transaction (L144 E8). However, the cases concerning the improvements to the irrigation system as well as discrepancies in the inventory made at the end of de Avila Romero's lease continued in the courts for some time, and finally the money was credited to his account with the Estate. Tomás de Avila Romero claimed to have some $60,000 worth of possessions in the 1740's (L144 E8), a sum that would count him prosperous. These lawsuits may have done him little damage, since in 1761 he was described as Asentista del Real Derecho de Alcabalas in Cuernavaca. His successor in the rental of Atlacomulco, Manuel de Olalde, had also been one of his bondsmen, owner of a store in Cuernavaca, and his predecessor as renter of the collection of alcabalas in Cuernavaca (L345 E19).

Manuel de Olalde was the last renter of the eighteenth century, holding a lease from December 1754 to December 1763 (L341 E10). Olalde offered $2,050 annually for a nine-year lease in 1754 (L341 E5), but shortly after he submitted this bid another was submitted by Domingo Joseph González del Pliego for $2,100. The latter was an Agente de Negocios, and after his bid was accepted he revealed that he had made it for Olalde. According to the terms of the contract, Olalde was bound to lease for the first five years, but the last four were optional. He told the Governor (L341 E14) at the end of the first five years that he did not wish to continue, but nevertheless did so and even applied for renewal of the lease in 1763 (L341 E10). The mill then was administered by the Estate for about seventy years, until perhaps 1833 (L402 E35).

Whoever leased Atlacomulco about 1833 found himself by 1837 ruined by the obligation (Alamán, *Documentos*, Vol. 4, pp. 379, 382) and unwilling to continue. It was leased in August 1837 to Juan de Goríbar and Anselmo Zurutuza (*ibid.*, p. 388). The former was co-owner with his brother of the plantation Cocoyoc, midway between Yautepec and Cuautla, and of the very old plantation Casasano, near Cuautla (Calderón de la Barca, pp. 396, 757). Anselmo Zurutuza was a Spanish friend of Calderón de la Barca's (*ibid.*, p. 752).

C. A RENTAL AGREEMENT

NOTE: The following is a translation of a rental agreement made between the Estate and the pueblo of Jiutepec.

In the pueblo of Xiutepec, subject to the villa of Cuernavaca, on the twenty-fourth day of the month of October of 1597, before the Licenciado Don Pedro de Losa Portocarrero, alcalde mayor of the Marquesado: there appeared Don Gaspar de Aquino, governor of said pueblo, and Dionysio Lázaro, mayordomo of the community, and Baltasar García and Juan Cortés, regidores, and Diego Cerón . . . yndios principales and naturales of said pueblo; and through Pedro de Nápoles, interpreter, stated that through him, as one, and in the voice and name of their community, have rented and did rent to the Marqués del Valle and to the jurado Lucas de Ocampo, mayordomo of the plantation of Tlaltenango in his name and here present, two pieces of land within the bounds of the estancia of San Francisco Zacualpa on one side and bounded by the river and the canefield owned by Luis de Casas, Alguasil Mayor, on the other; the one piece is called Acapan and the other Tepetenche, and both are and have been rented for the time and space of the next six consecutive years. Each piece of 20 brazas in width and 10 brazas in length, to the full amount of such pieces of land, is to be rented at 10 reales de oro común and all the value that said land earns, measured as stated above, is to be paid in cash at the beginning of each year, and this contract is to run from the first

of January next of 1598; and the owners of the land promise that during said time they will not repossess these lands unless they exchange for them others that are similar, as good, and in as good position and place, and for the same time and price. In the name of His Lordship, the aforementioned Lucas Ocampo agrees that he has accepted and does accept this lease on His Lordship's behalf, and after the pieces are measured will give and pay the first year all the amount owing, and when the first year ends will pay as much again; and this mode of payment he will keep to and comply with until the six years have ended; and he promises not to abandon the aforementioned two pieces of land during the stated time without paying for them even though unused, for fulfillment of which I [Lucas Ocampo] pledge the properties and income of His Lordship, and said Indians pledge their persons and properties, and the properties and income of their community.

Both parties give full power to the justices of His Majesty and His Lordship, of whatever place, to exert the full force of law and the power of execution to compel His Lordship and the mayordomo of the plantation, whoever he may be, to the fulfillment and payment of the aforesaid, as by sentence passed in adjudication; and they renounced the laws in their defense and in general, and said and granted that they asked and have asked that said alcalde mayor approve this contract and sanction it with his authority and judicial decree to the full extent of legal right; and said alcalde mayor, having seen this contract, stated that he does sanction and did sanction it to the extent of his authority and judicial power and signed it with his name, together with the contractants who knew how, being present as witnesses Padre Fray Francisco de Rixas, guardian of said pueblo, and Gregorio de Figueroa and Miguel Ximénez, being in said pueblo, and on account of and in partial payment of the first instalment of this lease they admit to having received 100 pesos de oro común. . . .

D. FIELD MEASUREMENT, SIZE, AND SHAPE

Units of Field Measurement. In many of the inventories, the description of average field size and typical field shape is not a wholly straightforward matter, since the units used to describe area differed at different times, and they were not in every case clearly defined. However, since a major feature of Chapter V is to describe changes in sugar yields over time, it is necessary to discuss the units of measurement and the problems they present. The linear measurements offer little trouble, because they are expressed in the conventional, widely used, and well-known units of vara and braza; it is instead the areal units called the tarea and the suerte, perhaps peculiar to the sugar industry, that offer difficulties.

The basic unit of linear measurement was the vara, sometimes used as a double length called the braza; for the vara I have used the English equivalent of 33 inches (Borah & Cooke, 1958, p. 11; Robelo, 1908, p. 18). In most inventories, as a first step, a vara was used to measure off a cord 50 varas long, a more practical instrument for field use than a single vara, and usually called until about the middle of the seventeenth century a mecate de cinquenta varas, later a cordel de cinquenta varas. The vara used when taking inventory to measure a mecate or cordel of 50 varas was usually specified to be a "vara castellana sellada del fiel contrato de la Ciudad de México" (1721), or more simply a vara (castellana) sellada. Since the canefields were the most valuable part of the inventory, and both parties were well represented at the inventory, it may safely be assumed that care was taken in the measurement of the fields. For this reason there was insistence on the use of the certified unit of length, and such care was taken that measurement of the canefields took three or four days to complete.

In the first set of inventories, from 1549 through 1607, the most commonly employed method of presenting the observations consisted of giving the dimensions of two sides (usually in brazas), thus implying that a rectilinear shape was common, or else the measurements were reduced to varas en cuadra or brazas en cuadra. The two methods of presentation are obviously related and appear rather straightforward, and are besides related to a third (1567, 1585) in which the measurements are given in suertes, defined in the following way by the surveyor (L432 E6): "que conforme a la medida, uso y costumbre usada y guardada en las islas de Canaria donde mas se practica lo susodho a de tener . . . cada suerte de 8,000 brazas, y cada braza de dos varas . . . conforme a la dicha medida de Canarias y esto es la q siempre he visto en las dhas islas por experiencia y medido por averme criado en yngenios de azucar y en la dha medida a mi leal saber no hay fraude." According to this statement, one suerte is approximately equal to 5.5 acres, and I have used this figure to convert the data of the inventories of 1567 and 1585 to acres.

In many inventories and accounts of the sixteenth century another unit of areal description, the tarea, was used in addition to the suerte. Its meaning is nowhere given, and it appears most properly to apply to the standard amount of cane assigned to be cut in a day rather than directly to a unit of area. The dimensions of the tarea are nevertheless the only clue to the amount of cane and sugar obtained per unit of area in the sixteenth and early seventeenth centuries, and the only clue to its dimensions is the approximate equivalence of 7 2/3 suertes and 46 tareas of cane given in 1568 in a document (L107 E3) concerned with frost damage to cane. It seems clear that here the suertes refer to the area of standing cane, whereas the tarea refers to an amount of cane that could be handled in one day by a mill, and is thus consistent with the general usage of tarea, in its sense of task, as is the fact that the early seventeenth-century accounts show that each tarea yielded generally 20 to 22 cartloads of cane. Since the suerte contained approximately 5.5 acres, the testimony of the experts in 1568 gives the tarea an area of 1.09 acres. I have used this figure in the absence of any other information concerning its size in the late sixteenth and early seventeenth centuries.

Fortunately, in the inventories of 1655 and 1671 the dimensions of all sides of the fields were given, stated in lengths of the mecate or cordel of 50 varas ("ordinaria sellada," 1671). This brief and rational interlude was replaced by a much longer one lasting through the inventory of 1721, in which the perimeter ("medida en contorno") only was given in varas. I have made use of the fuller dimensions given in the inventory of 1671 to draw to scale the fields

listed, assuming only that each field had one right angle to facilitate the drawing. Only two of the thirty-five fields formed a rectangle, and only eight others had two sides parallel; five were triangles. The usefulness of the inventory of 1671 lies in the fact that it provides an opportunity to estimate the area of an irregular figure whose sides are known, unlike other inventories that provide only perimeters. The relation of estimated field size to perimeter for 1671 is given by the expression $Y = 0.02X - 10$, where y gives field size in acres and X is length of perimeter in varas. There is no reason to assume that there was any important variation in field shape between 1671 and 1721, so I have used this equation to estimate the sizes of the fields until the later date, when a more precise system of areal description came into vogue.

The new method of describing areas that appeared in the inventory of 1754 persisted until the end of inventory records in 1884, but with changes in the dimensions of the basic unit of area, still called the tarea. Apparently one dimension of this rectangular form remained unchanged from at least 1763 until the Revolution of this century, because its width of 25 furrows remained approximately equal to 25 varas (L341 E10; Ruiz de Velasco, 1937, p. 202). The length of the tarea underwent startling changes, however, as can be seen in the tabulation below, growing from 13 varas (de apantle) in 1754 to a maximum of nearly 60 varas by about 1900, with the most rapid change coming in the last half of the nineteenth century. The areal changes were also large, from about one-twentieth of an acre in 1754 to slightly more than one-quarter acre by 1900.

Year	Length of Apantle, in Varas	Acre Equivalent of Tarea
1754	13.00 (some 14.00)	0.056
1763	14.00 (some 14.50)	0.061
1767	15.00	0.066
1777	15.00	0.066
1786	16.00	0.070
1791	17.00	0.074
1807	16.50	0.072
1811	18.00	0.078
1812	18.00	0.078
1847	25.00	0.107
1882	37.50	0.161
1900	47.76 or 59.70	0.200 or 0.257

Ruiz de Velasco (1937, p. 193) stated that a good-sized suerte consisted of 40 to 50 tareas, each of about 1,000 square meters, so by about 1900 the suerte (or 45 tareas) consisted of about 11 acres, just twice its size in the sixteenth century. This tarea of 1,000 square meters was probably used in plowing and weeding as the basic areal unit of piecework, and it differed from the tarea of irrigation (ibid., p. 218) of 632 square meters; Ruiz de Velasco does not explain why the difference existed. It is possible that the size of the suerte was approximately the same in the eighteenth century as it was when Ruiz de Velasco was working in Morelos, since a document of 1770 (L298 E50) states that a suerte should produce 1,000 loaves, or about 20,000 pounds of sugar, at a time when expected yields may have been about 2,000 pounds of sugar per acre. Unfortunately, I have found no eighteenth-century definition of suerte.

E. INSTRUCTIONS FOR ACCOUNTING AND OTHER PRACTICES

NOTE: The following is a translation of L257 E13, a document signed by the Marqués in 1588, containing instructions for the regulation of accounting and other practices at Tlaltenango.

Direction and order to be preserved in the expenses of the ingenio of Quernavaca, which are to be kept to as well by the administrator of my Estate as by the mayordomo of the ingenio and my controller, without any departure:

First, we must try to reduce the costs of wood, which are excessive, training mules and making oxcarts for the purpose of reducing this expense, and whatever is needed to do so shall be taken from my hacienda and charged to the account of the ingenio; and I wish to be told of all that is done, and whatever the administrator, mayordomo and controller believe is necessary to do, now as well as in the future, to diminish this expenditure.

For the expense of the table of Antonio Gaitán, now mayordomo, there shall be provided 300 pesos each year, one-quarter of mutton daily, with 8 pounds of beef, and no other expenditure shall be allowed.

All the stock slaughtered at the ingenio must be accounted for, and all must be weighed and rationed to Spaniards as to Indians, by pounds or aveldes; and this shall be as the administrator of my Estate has ordered, having told the mayordomo of the ingenio to do so, granting what each needs and no more, and the mayordomo may not depart from this rule except at his own expense.

The same rule is to be observed in the rations and disbursements of maize. All the maize necessary for rations and mules and for other purposes of the ingenio is to be released annually to the mayordomo, all at the same time, from the maize tributes of my Villa de Cuernavaca but not more or less than necessary and the maize is to be charged to the plantation account at one peso the fanega in the accounts of the administrator.

The mayordomo of the ingenio shall receive each year as salary one-twentieth of all the profit of the ingenio, minus the costs and expenses and losses of Negroes; and if this arrangement does not please him he shall receive the cash salary given to Martín de Espinosa, ex-mayordomo of the ingenio, and the ration mentioned above; and for the time that Antonio Gaitán is mayordomo he shall receive 100 pesos de minas more for expenses, and this amount is not to be exceeded, nor the 100 pesos de minas given to any other mayordomo; and my controller shall require Antonio Gaitán as soon as he becomes mayordomo to choose the plan that suits him, and the one he chooses shall be entered in the books of my counting-house, and from the day Antonio takes this position he shall enjoy his choice, and the account be drawn up accordingly.

He is also to account for the wine consumed and necessary for the ingenio, giving full details of its disposition, not merely the total as up to now has been done with respect of wine; and he is not to exceed this, nor shall the mayordomo be given more wine than the ration described above.

Once the administrator of my Estate, the controller, and the mayordomo of the ingenio have agreed upon the cattle — heifers and cows — necessary for the rations of the ingenio as described above, they must be at special pains to get the cattle from the estancia of Mazatepec as has always been done to meet these needs, and it is to be managed to supply cattle for the maintenance of the ingenio, so that it will not be necessary to buy cattle elsewhere, for when I was [in New Spain] it was customarily done in this way; and if it cannot be done so, and it is necessary to buy more cattle than are raised at the estancia, they must be brought from the estancias of Tuzcla and from those in Teguantepeque as the administrator thinks best, and I charge him to follow this rule.

In respect of the lease granted of the estancia of Atengo, where sheep are reared, it should be observed that it may be made only with condition that each year sheep must be provided for the ingenio at the expense of the lessee, as many as seem necessary for its maintenance, making it unnecessary to buy them, because, according to what I hear and what I saw when I was there, the rent of the estancia that is now paid is so low that it might well rise as much as the required sheep are worth; and their value is to be entered into the accounts of the administrator and charged at the same price to the ingenio.

In respect of extraordinary expenses, the mayordomo or the administrator shall be limited strictly, being told what they may spend on this account, and the amount shall not be exceeded, nor shall the controller admit any more in the accounts.

Each year without exception at the end of December and the beginning of January the mayordomo is to submit his account, and he shall receive no credits for goods or supplies without a corresponding reduction in their amount the following year.

The administrator, controller and mayordomo, after observing the rules that I have given concerning annual expenses at the ingenio, shall make a summary to send to me, telling me what they have decided and how much the expenditures at the ingenio shall be each year, in money as well as in stock, maize, wheat, and send it to me in very detailed form as soon as possible.

So that there shall also be order and account in respect of the molasses used at the ingenio, for it is necessary to consume and sell it, and we must know in what ways and why, they must tell me how much molasses drains from each lot of sugar that is made so that I understand fully; and this molasses shall remain to be sold by the mayordomo in the ingenio to cover [part of] its expenditures; and the rest of the money necessary for this purpose is to be granted by the administrator, as he is accustomed to do, so that the work shall not stop for lack of money; and I want the drafts made on Carlos Pérez or whoever sells the sugars in Mexico City, so that he may continue to supply money from the proceeds of the retail sales, in order that this hazienda continues to support itself with such sales, and if they are insufficient then let them still cover as much as possible, and the rest be given according to the opinion of the administrator.

I want stopped the salary given until now to the man who keeps the books and accounts [of the plantation], and appointed to the purgery a person who is suitable, to do what is necessary without more salary than we have been accustomed to give; and since in the accounts that are sent here to me from the administrator it is stated that some of the sugars called white in the ingenio accounts are nevertheless sold at a much lower price than are others because they are not so good — there never has been nor was in my time such a difference in price, and it does much harm — I order that this be corrected, and if the responsibility lies with the sugar-master, you shall find another who can do the work well, even if he must be paid more. If the losses due to this poor sugar cannot be explained in any other way than by the cane that is milled, it shall be thus explained to the administrator; in such a case the sugar shall be called low-grade by the mayordomo, and with the same manner and kind shall the seller of the sugars be charged, so that having seen the account of the mayordomo of the ingenio, and the administrator having inspected said sugars, it shall be decided how to charge the seller of them, so as to know which are good or bad and the price that each kind may get, and it should not be left in the hands of the seller of the sugar to call the sugar good or bad as he pleases, and which of the lots that he sells may be styled one or the other.

And this my instruction shall be kept as is described herein, without alteration, and if in some wise it be necessary to add or omit something, my administrator shall tell me of it, that I may decide what is convenient; and meanwhile all is to be observed to the letter; and I order that my controller shall apprise the administrator of my Estate and the mayordomo of the ingenio before a notary, sending me notice of the event; and this original instruction shall remain in my counting-house, my administrator and mayordomo to receive copies of it signed by my controller.
Dated in Sevilla, 14 June 1588.

El Marqués del Valle

F. FORM OF PLANTATION ACCOUNTS

DOUBTLESS from the time of establishment of the mill accounts were kept in much the same form as the earliest complete, available, annual account, which deals with 1581. Accounts earlier than this one exist only as fragments, and for some years in the 1540's there are available only a great many receipts representing purchases by the mayordomos. Although there is an Axomulco account covering a short period beginning in 1541, this is a copy (L277 E4) that may have been made in the late 1560's or early 1570's in connection with a suit brought by the heirs of the mayordomo of Axomulco against the Estate.

The accounts whose existence we may infer from the receipts of the 1540's, as well as those of the late sixteenth century, are primarily statements of the receipts and expenses of the mayordomo rather than of the plantation, and hence were checks on his honesty and performance rather than statements of profit and loss. This kind of account, generally supposed to be annual, was a simple collection of separate accounts comprising all the goods and money received plus goods produced, disbursed, and consumed; it included such categories as cash, foodstuffs (principally corn), clothing and cloth, livestock, cane planted, cut, and milled, and sugar made and sold or disposed of in other ways. Characteristically, each of the accounts begins with a statement of the amount on hand from the previous accounting period, receipts of the same good or cash, and disbursements. The carryover was added to the receipts, and from their sum was subtracted disbursements to obtain the balance carried forward, expressed as an amount owing by the mayordomo to the Estate, or by the Estate to the mayordomo. The annual accounts thus have the character of running inventories that divide the time from the assumption of the post of mayor-

domo to its renunciation; both of these events were marked by inventories as well.

That providing a check on the honesty and performance of the mayordomos was a major function of these accounts may be seen in the following analysis of the first account of Antonio Benítez Gaitán for the period from 4 November 1585 to 31 December 1586. The accounting occurred at the mill beginning 6 March 1587 in the presence of Juan Sánchez Adriano, chief accountant of the Estate; Gaitán was required to display his daily and weekly accounts as well as numerous receipts. The first statement in the account describes the period under review:

Mdlxxx vii a°s
En el yngenio de acucar que llaman Tlaltenango del Ills^mo s^r don martin cortes marques del valle que esta en terminos de su villa de Cuernavaca de la Nueva Espana a seis dias del mes de março de mill y qui°s y ochenta y siete a°s se fenecio quenta por Juan Sanchez Adriano Contador Mayor del Estado y hacienda de su s^r ills° con Ant° Benitez Gaitan administrador y mayordomo del dho yngenio de açucar de Tlaltenango de todo el açucar y otras cosas q an sido a su cargo desde cuatro de Noviembre de mill y qui°s y ochenta y çinco a°s que començo a administrarlo hasta fin del mes de Diçienbre de mill y qui°s y ochenta y seis a°s y para ella el dho Ant° Benitez Gaitan ysibio ciertos libros y otros rrecaudos por los q°s se le tomo y el dio la dicha quenta en la forma siguiente.

The account then moves directly to a discussion of the sugar made, and as with the other accounts bound together in this expediente, begins with the cargo to the mayordomo, derived from a daybook (libro) kept by Gaitán or under his supervision that showed the amount of cane milled (in numbers of carts) and the amount of sugar made. The production of each day on which milling occurred was listed, the date occupying a central column in which was written the numbers of carts of cane milled and the number of loaves of white, espumas, and respumas sugar made from the juice. To the left of this central column were two columns containing arabic numerals showing the numbers of carts and loaves of white sugar, to the right of the central column the numbers of espumas and respumas occupied separate columns. At the bottom of each page appeared the totals, also in arabic numerals, of each of the two outside columns, and at the end of the cargo of sugar the totals of all columns are written in words as well as in Roman numerals. The written summary is as follows:

Por manera que suma y monta el cargo de açucares que se haze al dho Antonio Gaytan como de suso se contiene nuebe mill y quatro cientos y noventa y cinco Panes y medio de acucar blanco y mill y qui°s y veynte y ocho panes de açucar despumas y quatrocientos y un panes de açucar de Panelas y seteçientos y noventa y dos panes y medio de açucar de Resp^as y el dho Antonio Gaitan açeto aprovo y dio por bueno y bien hecho el dicho cargo y juro en forma de derecho ques çierto y verdadero y que en el no ay fraude ni encubierta alguna contra su s^a ill^a que en todo el tienpo desta quenta no a cortado ni molido mas cana de las plantas y canaverales de su s^a ill^a de la de suso referida ni a molido ni benefiçiado otra cana alguna de particulares mas de la que de suso se contiene ni a rescevido ni entrado en su poder otra ninguna açucar perteneçiente a su s^a ill^a de que se deva azer cargo en esta quenta por ninguna bia ni manera hasta fin del dicho a° de qui°s y ochenta y seis y lo firmo de su nonbre

Antonio Gaytan

Following the information concerning the cargo of sugar appears the dacta (later "data"), arranged in the same way, with written information in a central column and flanked by separate columns listing number of loaves and their weight in arrobas; most of the sugar was sent to Mexico City to be sold, and the mayordomo typically presented at the time of accounting receipts from the storekeeper in Mexico City to whom the sugar was delivered. The total is given in Roman numerals, and the written summary of the data is similar to that of the cargo. This summary of the data was followed by a written description of the balance remaining after subtracting data from cargo; similar statements were made in sequence concerning the other two kinds of sugar, and the sugar account was then closed as follows:

y en la manera que dicha es se concluyo y feneçio la dicha quenta cargo y data y alcançes de açucar blanco y bajos y el dicho Antonio Benites Gaytan juro en forma de derecho ser çierta y verdadera y a verdado y entregado el açucar que da en data y descargo segun y como en esta qᵗᵃ se contiene sin que en ello ni en otra cosa de lo en ella contenido aya fraude ni encubierta alguna contra su sº illº asi aprovo y dio por buena la dha qᵗᵃ cargo y data alcançes y feneçimiento della y lo firmo de su nonbre

Antonio Gaytan

For the same time, cargo and descargo accounts were kept in the same way for cash, corn, sheep, cattle, cattle skins, cloth. In the case of the cash account both the cargo and descargo sections were signed by Gaitán, as the sugar account had been. Not all the supplies were treated in such detail as those listed above, since the Estate accountant merely relied on his knowledge of reasonable expenditures of supplies to check the amounts that the mayordomos claimed to have used. Instead, in this account there was included the following statement concerning the disbursement of supplies:

declaraçion de otras muchas cosas

y no se le haze cargo al dicho Antonio Gaitan del hierro açero cobre carbon vino vinagre azeyte pescado hava garvanço lenteja sal angeo çedaços tajamaniles çevo y otras cosas que por esta qᵗᵃ paresce aber conprado para el proveymiento del dicho yngenio porque el dicho Antonio Gaitan dijo y juro que todo ello y todo el çevo que a salido de las reses que se an muerto en el dicho yngº y las veynte arovas que le entrego Franᶜᵒ de Robledo todo ello se a consumido y gastado en el serviçio sustento y abiamiento del dicho yngenio y hazienda sin que dello aya quedado cosa alguna y particularmente el dho çevo se a gastado en untar los ejes de los carros y curenas y tenplas y husillos del yngenio y en alunbrar el molino todas las noches que muele y lo mesmo a sido de las demas generos y cosas de suso contenidas que todo ello se consumio y gasto en el sustento del dho yngenio y hazienda como dicho es [no signature].

There follows a short statement of the suertes of cane planted in the accounting period, and then a closing that, as was customary, contains two sections. The first states that both parties agreed to accept the account; the second is Gaitán's pledge to deliver the amounts of money and goods that he owed on balance. The solemnity of the occasion was emphasized by the presence and signatures of not only the Estate accountant, but also of the Governor, Francisco de Quintana Dueñas, and the Royal Notary of Cuernavaca, Pedro Martín de Málaga. Standard phrases found in, for example, rental contracts or contracts to deliver goods, occur at the end, indicating that Gaitán accepted the balances owing as equivalent to debts judged owing in court, and further pledged both his person and property toward their repayment. The four witnesses were Spaniards working at the plantation.

From the time of the oldest extant copy of an account set up in this way, in the 1580's, until about the middle of the

eighteenth century, there was little change in the form or purpose of the plantation account.

In 1770 Francisco Antonio de Larrea wrote from the Central Office of the Estate in Madrid to the Governor in Mexico describing how the plantation accounts were to be kept in the future, and directing him to forward a copy of the instruction to the administrator of the mill (L298 E50). The stated intentions of the changes in procedure and form were to provide better checks on the honesty of the employees and to increase the efficiency of the mill by requiring monthly rather than annual accounting. The instructions also required changes in the form of the accounts that were followed to the letter from this time until at least 1832.

The new form used the week as the basic accounting unit of time, and an example from the year 1785 is given in Figure 20. Each week's data and cargo were to be shown at the top of a fresh page in a book whose leaves were to be used consecutively throughout one year. Each weekly account bore a title similar to that shown in the figure, "Raya y Semana no. 1º. que comᵗᵃ. Domingo 2 y acava Sabado 8 de Eʳº de 1785." The data were placed first, with the several categories of day labor occupying the first place within the data. The names of the categories were placed in the column on the left, and under the next six column headings, one for each day of the week, were placed the numbers of day laborers in each category on each day. The eighth column gives the total in each of the rows, and the ninth the daily wage paid in each category. The tenth column gives the total amounts paid to each category of labor. Below the information concerning day labor appears salary payments to skilled labor and supervisors, and then the amounts paid for supplies and other expenses. The data were then summed, the total in the example being $655-1-0. Cargo appears next, with the individual amounts to the left of the column of totals; the sum of cargo is then compared with the sum of data and, in this example, the balance is stated thus: "Resultan a favor del administrᵒʳ. en esta Raya nº. 1º. la Cantidad de $519-0-6," which was then carried forward to the

Figure 20. Weekly account, 1–8 January 1785

117

following week, final adjustment occurring at the end of the year. Below the balance was placed the "Razón," or account of planting, milling, and other activities of the week; in this example it provides a statement of the number of loaves of sugar made, and also the weight of sugar sent to Mexico City in the same week.

De Larrea claimed that these changes would make accounting easier because, from the point of view of the administrator, they would lessen the onerous burden of an annual accounting at the end of each year; from the point of view of the Estate, handicapped by distance in its supervision, they would increase efficiency by giving the Governor in Mexico City a more nearly continuous check on performance. Also from the point of view of the Estate, the introduction of another change would decrease the likelihood of fraud: each of the three highest officials of the plantation — the administrator, the purgador, and the mayordomo, in charge of fieldworkers — was to sign each of the monthly statements, "concluyendo cada una antes de firmarla diciendo es cierta y verdadera y asi lo juramos a Dios y a esta santa [cross symbol] para que en todo tiempo sea valida." The administrator was directed to read the entire account, item by item, to the other two officials before they signed and the copy was sent to Mexico City.

G. MAYORDOMOS, DISPENSORES, AND PURGADORES

THE following sections include the names of mayordomos of the plantation during the periods when it was administered by the Estate and information about some of the dispensores and purgadores. Particularly in the case of the mayordomos, my purpose has been to convey a sense, however inadequate, of their activities, relationships, and prosperity. They occupied an economic and social stratum between the well-to-do and the artisan or tradesman class, and some were themselves owners or lessees of rural property. There must have been many haciendas, mines, ranches, and other economic enterprises in New Spain whose operation depended less on the direct attentions of the owners, many of whom were resident in Mexico City and might have owned or operated several rural properties, than on these people hired to attend to their day-by-day supervision. They are a very important but largely unknown class. I have included dispensores and purgadores because many of them were related to mayordomos or subsequently became mayordomos.

Names and Service of Mayordomos

The list below is as close to a complete account of the mayordomos as can be got from the records. Most of them had worked in other positions for the Estate, and many became administrator after serving as purger. The greater amount of information about sixteenth- and seventeenth-century administrators shows that many had relatives working at the plantation, but that was also at a time when many more Spaniards were employed than was the case later.

Name	From	To
Rodrigo Martín	?	22 Sep 1556
Martín de Espinosa	23 Sep 1556	31 Dec 1566
RENTED	1 Jan 1567	31 Dec 1580

Name	From	To
Bernaldino de Otalora	1 Jan 1581	27 Feb 1585
Francisco de Robledo	27 Feb 1585	28 Oct 1585
Antonio Benítez Gaitán	29 Oct 1585	21 Nov 1589
Cristóbal de Ribaguda Montoya	22 Nov 1589	9 Jy 1595
Lucas de Ocampo	10 Jy 1595	10 Apr 1606
Guillermo de Paz Cortés	11 Apr 1606	30 Sep 1607
Juan de Mendoza	23 Oct 1607	26 Sep 1609
Alonso Rodríguez Baza	27 Sep 1609	1 May 1611
Juan de Mendoza	2 May 1611	1 Feb 1613
Alvaro de Lugo y Montalvo	9 Feb 1613	17 Apr 1625
RENTED	18 Apr 1625	10 Jan 1644
Hipólito de Loya	11 Jan 1644	31 Dec 1645
RENTED	1 Jan 1646	31 Aug 1680
Francisco de la Salvatierra	1 Sep 1680	31 Dec 1682
RENTED	1 Jan 1683	24 Feb 1702
Diego Urtado	1 Mar 1702	1 Mar 1706
Cristóbal Mateos	2 Mar 1706	14 Aug 1709
RENTED (to Mateos)	15 Aug 1709	19 Aug 1718
Cristóbal de Bárcena	20 Aug 1718	20 Aug 1721
RENTED	21 Aug 1721	30 Je 1753
Agustín Fernández	1 Jy 1753	17 Dec 1754
RENTED	18 Dec 1754	10 Jan 1764
Joachim de Tovo y Latas	11 Jan 1764	2 Jan 1767
Andrés Vicente y Bitur	3 Jan 1767	18 Sep 1774
Michelena	19 Sep 1774	3 Nov 1774
Joseph Joachim de Urquijo	4 Nov 1774	6 Jan 1808
Manuel de Gaviña	7 Jan 1808	?
José Domínguez	?	1811
José Asencio Gordonís	1811	Sep 1818
José de Marón y Viñera	Sep 1818	23 Sep 1821
Juan Pio de Añibarro	23 Sep 1821	1826?
Lázaro del Corral	1826?	1829
Mariano Rubio	1829	30 May 1833
RENTED	30 May 1833	28 Oct 1847
Tomás de San Martín	1847?	13 Jy 1850
Aguirre	Jy 1850	185?
Ignacio Robles	187?	190?

Bernaldino de Otalora, first of the mayordomos concerning whom there is much information, was associated with the Estate and the plantation for at least twenty years, since he had been mayordomo at least as early as 1566 (L257 E13 D59), and apparently continued in the position after the plantation was leased to Miguel Rodríguez de Acevedo (L257 E13). He left the mill in 1571 (ibid.) but may have continued working for the Estate in some other capacity until he resumed the post of mayordomo in 1581. His death in February 1585 ended his association with the Estate. He was succeeded by an interim appointee, Francisco de Robledo, who may have been his nephew (L262 E1), and the testamentary executor of his estate was Ribaguda Montoya, himself mayordomo of Tlaltenango at a later date.

Francisco de Robledo was one of several persons having the same family name who were associated with the mill in the late sixteenth century. Otalora had given him power in 1583 (L257 E13) to act for him when he was absent, a power that Robledo found occasion to use in the following year (ibid.). Pedro de Robledo, possibly a relative, was paid by Francisco for work performed as dispenser in 1585 (L262 E1), a position of some responsibility. Antonio de Robledo's association with Tlaltenango began in 1582 and lasted until at least 1593; he worked for varying lengths of time under all the first four administrators in various positions: purger, cane supervisor, and dispenser. In addition, in 1591 and 1592, his mules carried sugar to Mexico City from Tlaltenango. Francisco de Robledo also owned mules that he rented to the plantation in the 1580's, and in addition he sold wood to the mill; in 1584 (L262 E1) he received $607 for wood sold to Tlaltenango, and in the following three years cane owned by him was processed at the mill on shares. Francisco de Robledo affords a good example of a person who occupied a variety of posts at the plantation, starting as cane supervisor in 1581, becoming purger in 1583 as well as acting as mayordomo for his uncle. In addition, he recruited Indian labor in 1585, for which purpose he took $80 and went "through the Marquesado to obtain Indians for periods of six days" (L257 E13 D3). Finally, he fell into disgrace in 1589, when Estate officials threatened him with imprisonment and confiscation of his property unless he produced receipts for expenditures amounting to $2,709 and for disbursements of sugar (L262 E1). After this year there is no mention of Francisco in the accounts, but Antonio de Robledo continued working as dispenser at Tlaltenango until 1591; after this date, following the misfortunes of Francisco, he was paid for carrying goods and sugar between Mexico City and the mill, and returned as dispenser for nine months in 1593 (L262 E2 D5). Thus, all three Robledos who worked at the plantation in the 1580's occupied positions of responsibility and one was briefly mayordomo.

The position of mayordomo was then held for slightly more than four years by Antonio Benítez Gaitán. There is little information available concerning him, but the little that exists suggests either a long life or a descendent of the same name, and the existence of relatives in the Cuernavaca region. He had been mayordomo of the Hospital de Jesús before becoming mayordomo of Tlaltenango, a sequence followed by his successor in both posts, Ribaguda de Montoya (L240 E1A). Either Antonio or Alonso Benítez Gaitán was corregidor of Cuernavaca about 1589 (L262 E5 D1). In the account of 1609–11 (L268 E2) appears the note that Alonso Benítez Gaitán sold 18 fanegas of maize to the plantation, and much later, in 1643, Alonso Benítez Gaitán registered title to land in the Marquesado, describing 6 caballerías (more than 600 acres) and one sitio de ganado menor on which he offered to pay a censo perpetuo of $400, including a license to grow cane. In 1588 (L262 E2 D7) Antonio Gaitán, as he was often called, milled cane on shares at Tlaltenango for himself and Alonso Gaitán, in spite of the Estate prohibition on such activity. Benítez Gaitán is an example of a mayordomo with local interest in the cane industry, with relatives in the Cuernavaca region, and a seeming prosperity.

His successor, Cristóbal de Ribaguda Montoya, was indeed active, prosperous, and a person with extensive local connections. He was in 1569 the son-in-law, heir, and testamentary executor of the estate of Martín de Espinosa, mayordomo of the plantation in the 1550's and 1560's and related to at least some of the many Espinosas who worked at both Tuxtla and Tlaltenango in the sixteenth century. In 1587 (L240 E1A) he acted as testamentary executor of the estate of Bernaldino de Otalora, when he held the position of mayordomo of the Hospital de Jesús, a post he left in 1589 to become mayordomo of Tlaltenango. However, his association with the Estate and status as vecino of Cuernavaca long predated this move; in 1576 he represented the Marqués in a lawsuit (L276 E77) and, even earlier, in 1566 (L228 E3) he witnessed the revocation by Juan Bautista Marín of the power that the latter had given to his nephew, Juan

Gerónimo Espinosa; in this document he was described as a vecino of Cuernavaca. From at least 1583 to at least 1603 he sold considerable amounts of wood to the mill, using his own mules to deliver it. Cane grown on his own account was processed on shares at the mill in fulfillment of a clause in his contract specifying that 12 tareas were to be accepted from him annually (L262 E5 D2); indeed, in 1592 (L262 E5 D3) most of the private cane milled at Tlaltenango was his. Ribaguda Montoya's contract, perhaps more favorable than most, also required that he was to receive 5 per cent of the profits of the mill. His documented period of service for the Estate was impressively long, extending nearly forty years from 1566 to 1603.

Lucas Ocampo's tenure as mayordomo was the second longest between 1581 and 1625, but unfortunately there is little information concerning his activities outside the time he held the post. However, he must have had long experience in the sugar industry, since Juan de Pastrama mentioned in 1605 (L263 E14) that he had known Ocampo for more than twenty years — that is, since about 1585; Pastrama was owner of the ingenio San Bernardo in Atlixco. Ocampo's first set of bondsmen contained the names of two Pachuca miners and Luis Suárez de Peralta, suggesting ties with Mexico City and the north rather than with tierra caliente. When a second set of bondsmen was required of him, the resulting list of five persons consisted of local notables in at least four cases: Diego de Benavides, Gregorio de Figueroa, and Diego Caballero must have been among the most important people in the Cuernavaca region, and all had long been associated with Tlaltenango; Lucas de León was at that time in charge of Estate sugar sales in Mexico City; the fifth, Salvador de Baeza, I have been unable to identify. Ocampo had, in other words, no difficulty in establishing ties with local notables after he assumed the position of mayordomo. His third set of bondsmen included again Salvador de Baeza and Diego Caballero, and in addition there appeared Gaspar de Ribadeneyra, a member of an important local family engaged in the sugar business.

Guillermo de Paz Cortés occupied the post of mayordomo for a relatively short time after Ocampo's resignation. I have no information concerning him beyond the fact that he was a nephew of Pedro de Paz (L28 V50 E17), encomendero by a grant from Hernán Cortés of the pueblo of Atotonilco south of Cuautla Amilpas. Pedro de Paz married Francisca Ferrer, doncella of the wife of Martín Cortés, and his two nephews, Guillermo and Antonio, shared in their uncle's property upon his death sometime in the 1570's.

Juan de Mendoza, successor to Guillermo de Paz Cortés, began his first tenure as mayordomo of Tlaltenango in 1607, but his earliest documented association with the mill occurred twenty years before, in 1587, when he was hired as bookkeeper at $200 annually. He worked full time as bookkeeper in both 1587 and 1588, and in 1589 assumed as well the post of purger at the mill (L262 E5 D4). In 1601 he reappeared as plantation doctor, a post he occupied through 1606, during and after which time he was also selling cane to the mill or having it processed there on shares (L268 E1, E2, E3). He became mayordomo in 1607, evidently as interim appointee, and was also described as "Mayordomo de los pobres del Hospital [de Jesús]" in 1607–8. His occupation of the post of mayordomo of Tlaltenango for nearly four years finished with another interim appointment to the position in May 1611, and the last notice I have found of him is contained in the order (L268 E4) from de Morga requiring him to account for money and supplies received by him in his last tenure, which had ended more than a year before.

There is virtually no information available concerning the mayordomo Baza, who occupied the post for a short time from 1609 to 1611. Similarly, I have no information concerning Alvaro de Lugo y Montalvo, in spite of his long tenure lasting slightly over twelve years and ending in 1625; however, it may be of interest to note that in 1644–45 one Gerónimo de Lugo, identified as a Spaniard, served as purger. Concerning the next two administrators, de Loya and de la Salvatierra, I have no information. Both Urtado and Cristobal Mateos have been dealt with in Appendix B (p. 112) since both leased the plantation and acted as its mayordomo for the Estate at different times.

Cristóbal de Bárcena assumed the post of mayordomo after the end of the lease of Mateos. He must have been well known to Estate officials, since he had acted for the Estate twenty-five years before he became mayordomo in the litigation concerning the Suárez Narciso lease of the plantation (L93 E8). This event occurred at the same time that one Joseph de Bárcena, also a vecino of Cuernavaca, assisted the Estate in some legal matters (L148 C3). In 1721 (L30 V57 E3) Cristóbal de Bárcena offered to lease the mill if no other person could be persuaded to take it off the hands of the Estate, noting that he could arrange with a friend in Mexico City to sell the sugar.

Following three more mayordomos—Tovo y Latas, Vicente y Bitur, and Michelena—about whom I have no information, began the administration of Joseph Joachim de Urquijo, noteworthy for its extreme length of thirty-two years and the apparently fully justified confidence that successive governors had in him. He retired in 1808 without any accusations of fraud, and his recommendation that Manuel Gaviña be his successor was accepted without reservation by the Junta at the meeting of 6 February 1808 (L219 E1). Urquijo offered to give help and advice to Gaviña, and further pointed out to the Junta that it would be difficult to find a good man for such "recio trabajo," especially since at any other hacienda a mayordomo would receive more pay than at Atlacomulco. The Junta decided, however, that Gaviña was to have the same salary as his predecessor.

Gaviña did not remain long in the post. He was succeeded by another short-term mayordomo, José Domínguez, apparently the poorest choice of the nineteenth century. An anonymous letter of April 1811 (L30 V53 E11) to the Governor claimed that Domínguez remitted molasses on credit to his father-in-law in Xochimilco and that it would be impossible to collect the value involved in the transaction; the writer added that in an effort to cover the debt, Domínguez had sent 600 cargas of maize—his own but raised at Atlacomulco—to Xochimilco for sale. Furthermore, he had placed an uncle of his wife's as storekeeper at Atlacomulco and charged his own food costs to the store at the rate of $9 or $10 weekly. The Estate lawyer advised the Governor that there was no need to start suit on the basis of an anonymous letter, advising the latter to proceed as any other "padre de familia diligente"—that is, by sending the Alcalde Mayor of Cuernavaca to the plantation to look for a shortage, and to put Domínguez in prison if one were found. Another anonymous letter from Xochimilco followed in May, describing the rise of the Domínguez family on their ill-gotten gains:

"Aqui es publico y notorio que robaron todos los que estuvieron en aquella hacienda perteneciente a Dominguez, tienen sus bienecitos, ellos estaban enseñando las carnes porque no habia con que taparlas, ahora parecen Marqueses todos. . . ." I have found no document stating how these charges were dealt with, but Domínguez was replaced very shortly.

His successor was José Asencio Gordonís, who served for seven years and then resigned to marry in Mexico City, where he became administrator of the houses belonging to the Estate; he was described in 1821 as Capitán retirado and a businessman of Cuernavaca (L244 E30), so it is apparent that his record was satisfactory. Gordonís was succeeded by José de Marón y Viñera in 1818 on the former's recommendation, which included a statement concerning the importance of his ability and contacts, as well as the information that he had served both at Atlacomulco and other haciendas in the region for a long time (L219 E1). Marón had been purger at Atlacomulco since at least 1812, and was killed in March in action against Iturbide and Guerrero (L219 E2). His death caused the Estate controller some concern, inasmuch as it had become customary for the assistant administrator of the plantation to become mayordomo. At this time the controller's son, Don Vicente Ramírez, was assistant to the mayordomo, and hence in line for the principal position, but the controller agreed that he was too young for such responsibility and lacked the experience of Marón.

The post of mayordomo was then assumed by Juan de Añibarro, who had succeeded Marón as purger at Atlacomulco when the latter was promoted to mayordomo; Gordonís had recommended Añibarro for purger on account of his age, good health, and "other good qualities" (ibid.), having known him for some years and employed him to manage Gordonís' own affairs in the store at Atlacomulco. Concerning the next mayordomo, Lázaro del Corral, I have no information. Mariano Rubio, his successor, remained in the post until the plantation was rented in 1833, then may have continued in the service of the Estate in Mexico City (L216 E1); apparently his tenure was satisfactory. Tomás de San Martín assumed the post of administrator on the return of the plantation to the Estate in 1847, and died in office in the cholera epidemic of 1850, as did the purger, Alamán (Documentos, Vol. 4, p. 491) described him as a very good mayordomo, and in fact the Estate owed him money for operating expenses at the time of his death (ibid., p. 637).

Alamán appointed Aguirre as successor to San Martín, in spite of his tender age of twenty years, describing him two years later as a little too "complaciente" (ibid., p. 608).

No information is available concerning administrators who may have been present between the tenure of Aguirre and Ignacio Robles. The latter occupied the post at least by 1878 (L420 E14), and a few of his letters to Juan Bautista Alamán, son of Lucas Alamán and his successor as manager of the Marqués' affairs in Mexico, still survive. Most of those I have seen are in L420. The latest is dated 3 December 1905, and signed by Robles in a hand even shakier than that in a letter of 1891 (L420 E15). As with the tenure of Urquijo in the late eighteenth century, so the tenure of Robles extended for a very long time, and he appeared to have the complete confidence of the Marqués' agents in Mexico. It would be interesting to know if Ignacio Robles was related to Lorenzo Robles, vecino and businessman of Cuernavaca, who cashed libranzas of Atlacomulco in the 1830's (L457), was a correspondent of Lucas Alamán's, and, according to the Fishers (p. 752), administrator of Atlacomulco in 1841 for the renters Goríbar and Zurutuza.

Dispensores and Purgadores

Besides Antonio and Francisco de Robledo, there is information concerning six other persons who occupied the post of either purger or dispenser in the period 1581–1625. It is possible that another dispenser, Pedro de Robledo, was related to the other persons of the same name, particularly since their periods of association with the plantation overlapped. In addition, Gerónimo de Montalvo worked at the mill in the last year that Alvaro de Lugo y Montalvo served as mayordomo, and Juan Bautista Tenorio was dispenser and purger successively in 1623 and 1624 (L269 E34; E36), just before the plantation was leased to Andrés Arias Tenorio.

Six of the dispensers other than the de Robledos worked for long times at the mill in various positions. Three either sold cane to the mill or else had some of their cane milled there on shares; one of them, Cristóbal Manzano, was identified as a resident of Cuernavaca (L268 E3), and another, Sebastian Sarmiento, was vecino of the same city (L262 E5 D4). In the course of his long association with the mill, extending from at least 1588 (L262 E2 D7) to at least 1624 (L268 E36), the latter sold wood to the mill, carried sugar on his own mules to Mexico City (L262 E5 D3; D4), was associated in 1606–7 with Juana de Espinosa, perhaps as overseer, bringing wood to Tlaltenango on her mules (L268 E1); in 1607–9 he was teniente to the Corregidor of Cuernavaca, representing the latter at an inventory of the plantation (L268 E3). Gerónimo de Espinosa was purger from 1599 until 1604, working full time throughout the years in that capacity; in addition to this rather unusual steadiness, he sold wood to the mill in 1600, 1601, 1602, and 1603 (L263 E15; E13), the values amounting to slightly more than $1,000 annually. Possibly he was related to some of the other Espinosas who had been earlier associated with the mill, owned mules, and worked long periods for the Marqués. Antonio Izquierdo stated in 1605 (L263 E14) that he had worked at Tlaltenango for five years "about fifteen years before." There is no record of such service beyond identification of him as purger in 1587 (L262 E2 D5), but in subsequent years he carted sugar to Mexico City for the mill on his mules, had his cane processed on shares, and cashed drafts of the Governor of the Estate in Mexico City in 1603 (L263 E12). Obviously, he had been in the region for a long time, since he stated in 1605 that he had known Lucas Ocampo for more than thirty-five years, or since about 1570 (L263 E14). Miguel Ximénez worked for the mill intermittently for at least twenty years, from 1587 (L262 E2 D5), when he was hired for three months as muleteer, until 1606, when he was dispenser; in addition he sold wood to the mill, and sold meat to it in 1607 (L268 E3), when he may have been responsible for the meat supply of Cuernavaca. In 1608 (L268 E3) he was the testamentary executor of the estate of Gaspar Martín, another long-time employee at Tlaltenango. The intermittent association of Mateo García Perdomo with Tlaltenango extended from 1613 to 1625; he was dispenser several times, his second occupation of the post ending when he fell sick.

H. DIVERSIFICATION OF PRODUCTION IN THE NINETEENTH CENTURY

THE unrest and disturbances associated with the Independence movement, in which the life of a mayordomo had been lost, the associated difficulties of getting workers, and the low prices of sugar and molasses, caused the Junta of the Estate considerable concern, expressed in the minutes of their meetings of 1821 and 1822. In August 1822 the mayordomo Añivarro wrote that the cane was in poor condition, lodged, touched by frost, damaged by rats, and suitable only for making molasses; in addition, the stock was being stolen (L219 E2). The Governor Fernando Lucchesi sought to rescue as much as possible from this unpromising situation, complicated by the fact that in 1822 the Marqués had been ordered by the Government not to rent or sell the plantation (ibid., minutes of 8 July 1822). At the meeting of 4 April 1823, Lucchesi suggested that it might be profitable to make rum and aguardiente at Atlacomulco by means of a "recently invented steam device" not described in detail, and he also hired one Frederick Waulthier, a Frenchman, to make a survey of the plantation and suggest a plan for diversifying its production in order to lessen the dependence of the Estate on sugar and molasses.

Waulthier reported to the Junta at their meeting of 12 May 1823, outlining his plan nearly to eliminate the growing of cane at Atlacomulco, substituting coffee and indigo; cacao was rejected as a possibility, as Alamán later explained (ibid., minutes of 12 April 1825), because the land was too dry. Coffee had been planted at Atlacomulco in 1805 and 1806 and began to yield in 1812 (ibid., minutes of 1825), so that its suitability had already been tested at the hacienda. Indigo had apparently interested other local hacienda owners, and at the Junta meeting of 7 August 1824 (ibid.) Don Manuel de Eguía was reported to have written from Spain that Cuernavaca indigo sold in Spain for a higher price than Guatemalan because of the superiority of its dye. Henry Ward (Vol. 2, p. 77) regarded indigo as an experimental venture at the time of his visit to the region in the 1820's, writing that planters were trying to introduce its cultivation into Cuautla. Apparently diversification of agricultural production was being widely considered at that time in Morelos.

Waulthier's plan was agreed to in part by the Junta and Governor at the meeting of 22 May 1823, including the provision that sufficient cane should be retained after the next milling to plant 1,200 tareas; this would be enough cane to keep the milling and processing machinery in order in case the price of sugar should rise. In fact, it did rise sufficiently within a year to cause the planting to be enlarged to 1,600 tareas, but the Governor finally ordered the planting of 1,900 and the mayordomo increased the figure on his own initiative to approximately 2,100 by December 1824.

The Junta and Governor also agreed in 1823 to sell all the mules in excess of 200 and buy cows in order to be able to offer milk and veal for sale locally. Waulthier then departed for the plantation to supervise the changeover and arrange for the rental of the nearby indigo processing facilities of Don Simón de Castro. He also arranged to visit nearby coffee plantings in order to study their management. The Junta and Governor were particularly interested in reducing the labor requirements of the plantation, intending finally to depend only on the gente of Atlacomulco and to avoid the need for recruiting Indians from surrounding pueblos. Their hope was that diversification would contribute to the reduction of labor requirements.

Waulthier reported at the meeting of 23 June 1823 that he could rent the Castro indigo facilities for $1 per day, and bought 6 cargas of indigo at $75 each to begin the operation. After visiting the coffee plantings at Cocoyoc he decided that the plantains he had previously thought necessary to shade the coffee seedlings, which would also have resulted in yet a different crop for the plantation, were not in fact necessary, laths being preferable. He proceeded to buy the first seed in order to begin planting coffee in September. Between 85,000 and 90,000 seeds were planted, but by March 1824 only 50,000 had sprouted; even this figure was higher than Waulthier's original suggestion of 30,000 or 40,000 coffee trees. Apparently some replanting was necessary because some seedling died after transplanting, with the result that the trees by January 1825 were of uneven ages, making it difficult to provide proper shade. The second planting from which the new stock had come was made in September 1824; germination was poor because the seed was bad, according to the mayordomo, and heavy rains caused much of the seed to rot before germination. By this time the administrator Añivarro was on his own, depending on other local persons for advice and assistance, or on consultations with Waulthier in Mexico City, for the latter was then too busy to return to Atlacomulco.

The indigo planting must have been successful, since there was sufficient indigo to require the use of Castro's mill in 1824 and 1825. However, it was reported by the administrator early in 1824 that the mill was so far from the fields at Atlacomulco that it was too costly to carry the indigo to it, and he suggested putting processing works on the north side of the sugar mill. In November the administrator was asked to draw a plan of the mill showing where the indigo and coffee works should be (his sketch of 1824 provided the basis of Fig. 9, p. 51) and with the proposed additions his plan describes fairly accurately the plan of the eastern half of the existing ruins of the mill.

The two inventories of 1829 and 1884 contain some information concerning the coffee plantings. Five fields were listed by the names they had borne since about the mid-eighteenth century. The names, together with the numbers of trees they had in 1829, are as follows:

	Number of Plants	Condition
La Soledad	10,556	Good; bearing
San Juan	12,285	ca. 6 per cent poor; bearing
Poder de Dios	11,996	ca. 4 per cent poor; bearing
San Vicente	10,742	Poor
San Dimas	6,924	Adequate; most plants a little yellow

The oldest and largest plants were in La Soledad, between 6 and 8 feet high, whereas those in San Dimas were less than 3 feet high. The first four fields were all enclosed by orange trees, 525 in number and about 12 feet high, and the entire planting by a fence of lemons that were only slightly smaller, for a total length of approximately 2,850 yards. By 1884 the experiment with coffee had lost its importance, since there were only 11,513 coffee plants, or enough to occupy the equivalent of only one of the first four fields. One or two of the others, however, must have been in plantains and bananas—"platano, costarrica y guineo"—and most of the oranges had given way to Manila mangos, almacates, zapotes, membrillas, peaches and apples; the lemon trees remained on only two sides of the coffee trees.

Fanny Calderón de la Barca (pp. 375–76) gave a vivid description of the coffee grove and orange walk at Atlacomulco in February 1841, concluding with the remark, "We felt this morning as if Atlacomulco was an earthly paradise." In fact, however, San José de Cocoyoc turned out to approximate this state more closely, since its orchards of three thousand orange trees and an enormously greater variety of fruit trees than Atlacomulco moved the traveler to put the matter more simply: "I never saw a more beautiful sight" (pp. 396–97). Ample precedent for raising oranges for sale was present in Yautepec, which Ward (Vol. 2, p. 303) described as a very pleasant sight because of the abundance of orange trees it contained; the riches of the inhabitants, he wrote, were these groves, from which they supplied Mexico City and Puebla. It does not appear that the vogue for such elaborate plantings had reached all the haciendas of tierra caliente, since Señora Calderón de la Barca did not remark on such groves in the other plantations she visited. It was, of course, to Cocoyoc that Waulthier had traveled to obtain information on the cultivation of coffee, and at the time of the visit of Calderón de la Barca its owner was also renting Atlacomulco.

Señora Calderón de la Barca stated (p. 377) that the coffee plantation contained "upwards of fifty thousand young plants, all fresh and vigorous," so the grove had not been extended beyond the plantings of the mid-1820's.

Probably the plantation was less profitable than beautiful. In the three years 1828, 1831, and 1832, the production of coffee amounted, respectively, to 15,000, 16,000, and 18,000 pounds. Alamán wrote to the Marqués in September 1837 (Documentos, Vol. 4, p. 389) that the cafetal probably produced less profit per acre than cane, and repeated his statement slightly more than a year later (ibid., p. 418), at a time when he felt that political conditions in Mexico were such that if anyone offered $100,000 for Atlacomulco the offer should be accepted immediately. He also felt that the cafetal added sufficient value to the hacienda to justify lowering the rent in order to make certain that a lessee would not only maintain the plantings but also improve them (ibid., p. 423). Goríbar and Zurutuza fulfilled their part of the next lease contract, renewing most of the planting, so that Señora Calderón de la Barca's description of the plants as "young" in 1841 was accurate; in other words, the plantings of the 1820's lasted little more than a decade before requiring replacement. However, the cost of replanting was approximately $6,000 (ibid., p. 489) and the renters were to receive the heavy March crop of 1848, after the plantation had been returned to the Estate in the previous year. This arrangement had been made in lieu of interest payments on the $6,000 and the cost of other improvements made by the renters, which the parties agreed should be repaid in instalments finished at the end of 1849 (ibid., p. 462).

Alamán expected that the proceeds from the sale of plantains interplanted with the coffee and the oranges that were also part of the cafetal would cover the costs of production of the coffee (ibid., p. 489). In the crop year 1847–48, when the crop was to go to the lessees, the oranges and plantains did not cover production costs; in 1848–49, when the crop was poor, there was a profit, but in the following crop year, owing to the death of the administrator San Martín and complications concerning his accounts, Alamán was not certain there would be a profit. The success, if it occurred, was probably not spectacular because Alamán did not mention it. In the crop year 1851–52 Alamán expected little because the plant yielded lightly in alternate years; this implies that the previous year's yield had been satisfactory (*ibid.*, p. 559). In fact, by mid-August of 1852 (*ibid.*, p. 636) the cafetal had lost $426, and he wrote in the following January that, as he had expected, little had been produced, but the trees were flowering heavily (*ibid.*, p. 653).

Alamán's optimism concerning the cafetal continued in spite of its rather poor showing and its poor condition at the beginning of 1853 (*ibid.*, pp. 653–54). On account of the age of the trees and the fact that there were many empty spaces, it appeared to him that only about half the area of the cafetal was producing, but he hoped that putting new plants in the empty spaces and cutting back the older plants would improve its productivity. He planned on 20,000 new trees, as well as replacements for some of the oranges and lemons: "Y con esto y otros arreglos si Dios me da, no tantos años de vida como los 103 del difunto tío Tomás, sino tres o cuatro mas, me prometo ver el cafetal en estado de buen producir."

He had been ill for some time, writing in 1850 that perhaps he would improve if he spent at least part of the winter at Atlacomulco. He died on 2 June 1853, without realizing his plans to improve the cafetal.

TABLES, GLOSSARY, AND BIBLIOGRAPHY

Tables

NOTE: The following tables have been compiled primarily from accounts of the plantation and the Estate, as well as inventories of the plantation. Omitted data reflected missing or incomplete records. In the case of plantation records, the sources are given in the bibliography. Other sources of data are given in the tables or in the text.

Table 1. Changes in Estate Income (in Pesos)

Source	ca.1570[a]	1681	1770	1807	1823	ca.1850
Tributes. . . .	47,084	15,839	31,013	45,365	n.a.	0
Censos.	0?	2,875	8,454	16,753	7,255	5,000[b]
Haciendas Marquesanas[c] . .	5,000[d]	2,700[e]	2,000[e]	3,000[e]	2,600[e]	0
Ingenio Atlacomulco	9,000[e]	3,666	2,100[f]	21,251[g]	-8,000[h]	30,000[i]
Ingenio Tuxtla.	5,000[c]	0	0	0	0	0
Tehuantepec . .	3,667	5,225	5,399	5,399	0[j]	0
Peñol de Xico[e].	150[d]	140	366	305	350	0?
Houses, Mexico City.	3,035	7,239	2,207	28,922	24,658	n.a.
Plaza del Volador.	0?	70	2,400	3,000	3,000	0[k]
Other urban . .	0?	n.a.	1,782	1,732	899	0?
Miscellaneous .	2,105	10,753	n.a.	3,159	450	0?
Taxco mines . .	4,764	0	0	0	0	0
Arrears collected. . . .	n.a.	4,576	n.a.	n.a.	n.a.	n.a.
Gross income	113,179	58,083	55,721	128,887	31,212	n.a.
Expenses . .	n.a.	n.a.	26,761	n.a.	n.a.	n.a.
Net income .	n.a.	n.a.	28,960	n.a.	n.a.	n.a.

[a]This column is derived from the voluminous records (L226 E2) kept from 1567 to 1574, when the Estate was under the control of the Crown. As in the rest of the Estate accounts, amounts due in any category, whether collected or not, were listed as cargo (in this table, they contributed to total gross or expected income), and should be deducted as "expenses" in this table (data or descargo in the originals) if written off as uncollectible, to obtain net income; unfortunately, it was not possible to find expenses, including expenses of collection, for either 1570 or 1807, so the figures given do not represent net income. Nor have I attempted to estimate the long-term average collected by the Estate on amounts owing; it might be reasonable to assume that it was less than 75 per cent of the amounts expected or due. [b]Average, 1844-51. [c]The rural properties in Oaxaca and Tehuantepec. [d]Estimate. [e]Rent. [f]Income if rented, 1764-69 (L36 V67 E12). [g]This figure appears high; it probably refers to value of production rather than net income, which in 1809 was about $4,000. [h]Average annual loss, 1817-21. [i]Average, 1848-52. [j]Payment suspended, 1 January 1814. [k]Sold to Ayuntamiento of Mexico City.

Table 2. Estate Income Expected and Sent to Spain (in Pesos)

Year	Amount Expected	Amount Sent
1619.	n.a.	21,497
1629-37[a]	n.a.	44,658
1633-39	48,000	n.a.
1639-48	51,000	30,765[b]
1640.	n.a.	35,877
1643.	n.a.	34,570
1647.	n.a.	15,824
1649-58	38,000	n.a.
1661.	38,000	n.a.
1669.	36,000	36,773
1671.	38,100	16,186

[a]Average annual shipment; 1633 missing.
[b]Average annual shipment; owing to depopulation, the rent was readjusted in 1645 (L237 E17).

Table 3. Summary of Governor's Accounts, 1686-1704 (in Pesos)

Year	Received	Disbursed	Balance	Atlacomulco
1686.	45,009	4,340	40,669	4,000[a]
1687.	59,719	6,551	53,168	4,000[a]
1688.	54,322	5,201	49,121	4,000[a]
1689.	47,478	4,943	42,535	4,500[a]
1690.	54,236	3,941	50,295	4,500[a]
1691.	53,503	4,180	49,323	627
1692.	44,436	4,965	39,471	-23,580
1693.	54,361	5,698	48,663	3,000[a]
1694.	49,494	23,533	25,961	3,000[a]
1695.	59,742	28,947	30,795	3,000[a]
1696.	43,178	6,427	36,751	3,000[a]
1697[b]	12,533	3,175	9,358	2,000[a]
1698[c]	89,302	12,874	76,428	4,000[a]
1699.	50,526	16,239	34,287	3,000[a]
1700.	43,581	7,905	35,676	3,000[a]
1701.	53,148	12,457	40,691	3,000[a]
1702.	46,807	27,210	19,597	1,797
1703.	51,413	8,972	42,441	1,797
1704.	54,641	7,587	47,954	1,797
Average. . .	50,917	10,271	40,647	1,602

[a]Contracted rent. [b]9 months. [c]15 months.

Table 4. Payments According to the Will of Hernán Cortés (in Pesos)

Item	Single Payments	Annual Payments
Expenses of funeral.	5,000[a]	0
Hospital, convent, and institute . .	360	5,500
Dowries and gifts.	300,000	0
Miscellaneous annual pensions. . . .	0	2,800
Annual payment to Juana de Zúñiga. .	0	11,000
To guardians of Martín Cortés. . . .	12,000	0
Income of Martín Cortés to age 20. .	0	16,500
Debts of Hernán Cortés	0[b]	...
Total	317,360	35,800

[a]Estimated. [b]The will does not indicate their total amount, but it seems not to have been large.

Table 5. Amounts Owing by Governors (in Pesos)

Governor	Period	Amount	Source
Pérez de Ayala.....	1580's	>180,000	L246 E1
Santa Cruz........	1590?-1604	...a	L245 E14
Leardo..........	1605-14	>170,000	L246 E1
Carrillo y Alarcón..	1618-46?	...a	L325 E3
Valles...........	1649-58?	...a	L328 E2; E10
Ortega...........	1658?-60	...a	L346 E28; E31 L320 E1
Ruiz de Colina.....	1660-70	large amount	L346 E3; E14 L238 E1
Avello and Sandoval..	1671-75?	large amount	L317 E13; E19
Fajardo de Aguilar..	1677-80?	...a	L317 E4; E5
Esteban de Yturbide..	1730-38	80,000	L36 V67 E5
Bermúdez Sotomayor..	1739-43?	large amount	L140; L144
Vértiz...........	1748-53	8,066	L1 V3 E13

aAmount owing is indeterminate.

Table 6. Profits and Losses at Atlacomulco, 1702-60 (in Pesos)

Manager	Period	Profit	Loss
Administration....	25 Feb 1702 to 31 Dec 1704		5,391
Administration....	1 Jan 1705 to 28 Feb 1707		3,922
Administration....	1 Mar 1707 to 31 Jy 1708	4,561	
Administration....	1 Aug 1708 to 14 Aug 1709	3,379	
Total.........		7,940	9,313
Balance........			1,373
Rented to Mateos..	14 Aug 1709 to 20 Aug 1718		17,999
Rented to Palacios..	20 Aug 1721 to 20 Aug 1731		1,385
Rented to Matajudios.	20 Aug 1731 to 6 Sep 1741		10,633
Rented to Romero..	6 Sep 1741 to 30 Je 1753		12,450
Administration....	7 Je 1753 to 17 Dec 1754		8,015
Rented to Olalde..	17 Dec 1754 to Dec 1760		10,885
Total.........		7,940	62,740
Balance			54,800

Table 7. Sources of Cash for Atlacomulco (in Pesos)

Year	Libranzas Total No.	Total Persons	Total Value	Mean Value	Governor	Tributes	Sales of Molasses and Sugar	Approximate Totals	Total Plantation Expensesa
1718...	11	6	7,023	1,170	0	7,532	2,938	17,493	23,011
1719...	12	8	4,080	510	2,310	5,368	2,510	14,268	18,165
1720...	17	10	5,440	544	3,900	2,800	2,226	14,366	20,930
1768...	30	9	7,320	264	0	0	4,166	18,335	19,661
1769...	16	7	4,210	n.a.	n.a.	n.a.	n.a.	n.a.	n.a.
1785...	28	13	23,386	1,800	n.a.	0	7,048	n.a.	29,965
1811...	5	3	4,825	965	n.a.	0	n.a.	n.a.	n.a.
1819...	27	7	24,832	3,547	n.a.	0	15,954	n.a.	n.a.
1822...	24	9	19,682	2,187	n.a.	0	16,638	n.a.	37,333
1828...	37	13	17,248	1,326	n.a.	0	n.a.	n.a.	n.a.

aTheir total data.

Table 8. Expenses of Estate Sugar Sales in Mexico City (in Pesos)

Year	Source	Salary	Food	Loss Allowed	Commission	Minor Costs	Rent
1590's	.L241 E2						
1680-81	.L148 C1						60
1702-9.	.L30 V56 E1			3%			
	L30 V58 E1	350	150	3%			120
1770's	.L341 E4						
	L231 E5	350	150	3%	4%		
1779.	.L232 E3	1,000a		allowed			294
1785.	.L232 E101	300	200	2%			294
1818.	.L298 E28	600a		3%			294
1825.	.L152 E18	300	?	allowed		225	294

aIncludes food allowance.

Table 9. Ownership of Indian Land Rented for Tlaltenango, 1549

Owning Barrio	Number of Pieces	Acres	Annual Rent, Pesos	Rent per Acre, Reales
Rented Land to be Returned				
Olaque.........	9	111.0	98-3-0	7
Tequepaque........	4	65.4	72-0-6	9
Panchimalco.......	7	51.7	54-0-0	8
Tetela.........	4	25.9	32-2-0	10
Cocosingo.......	3	16.6	12-1-6	6
Xala..........	2	1.1	1-1-8	9
Private?........	1	0.5	0-4-6	9
Other--Tetela.....	1	n.a.	(9-2-0)	...
Total.........	30	272.7	273-3-8	
Land to Continue in Rental				
Not identified....	...	121.5?	70-0-0	...
Olaque.........	3	38.5	32-4-0	7
Tequepaque.......	1	30.5	32-0-0	8
Cocosingo (millsite).	1	13.3	11-6-0	7
Total.........	5	203.8?	146-2-0	

Table 10. Summary of Rent Paid to Indians

Year	Amount	Year	Amount	Year	Amount	Year	Amount
1581[a] . . .	$50-7-0	1593 . . .	188-6-6	1600 . . .	157-4-0	1621 . . .	179-2-0
1583[a] . . .	75-2-0	1594 . . .	255-4-0	1601[e] . .	382-4-0	1622 . . .	158-0-0
1584[b] . . .	259-0-0	1595 . . .	191-2-0	1602 . . .	10-4-0	1623 . . .	109-0-0
1585-86[c] .	356-5-6	1596 . . .	182-2-6	1603 . . .	178-0-0	1624 . . .	581-6-0
1586-87. .	398-3-0	1597 . . .	216-6-6	1606-7 . .	385-6-6	Total. .	2,922-0-0
1588 . . .	253-7-0	1598 . . .	120-6-0	1610-11. .	98-0-0	Average .	182-0-0
1589[d] . . .	153-1-0	1599 . . .	106-6-0	1611-13. .	124-4-0	1718-21[f] .	165-0-0
1591 . . .	202-0-6	Total. .	3,252-0-0	1613-15. .	120-0-0	1785[g]. . .	150-0-0
1592 . . .	241-4-6	Average. .	203-0-0	1620 . . .	437-0-0	1831[h]. . .	200-0-0

[a] Much owned by individuals. [b] Most in Tlacomulco. [c] Paid to 37 individuals. [d] Account states paid only to Indians. [e] Most rented for 2 years. [f] Rent was for pasture at about $55 per year. [g] Rent of Acatlán to Cuernavaca. [h] Rent of Acatlán and la Huerta.

Table 11. Area in Cane, Types of Cane, and Field Size at Atlacomulco

Date	No. of Fields	Acres in Cane and Plowed	Tareas	Average Field Size, Acres	Median Field Size, Acres	Range in Field Size, Acres Min.	Max.	Plant Cane, Acres	Ratoon Cane, Acres	Plowed Land, Acres
Jy 1549	24[a]	274	...	11.4	7.9	0.8	95.0	...	...	32
Jan 1567. . . .	...	217[b]	...	...	...	...	...	130	86	0
Feb 1577. . . .	...	220[b]	...	...	...	...	...	118	102	0
Nov 1585. . . .	20	269[c]	...	...	...	...	...	n.a.	119	0
Feb 1590. . . .	34	403	...	11.8	8.1	0.3	66.4	172	231	0
May 1606. . . .	17	112	...	6.5	5.6	0.8	18.2	...	...	...
Oct 1607. . . .	25	200	...	8.0	4.1	0.7	19.4	96	104[d]	...
May 1611. . . .	...	...	114	...	...	...	...	50[e]	64[e]	...
Feb 1613. . . .	...	...	64½	...	...	...	...	54[e]	10½[e]	...
Dec 1618. . . .	...	...	122	...	...	...	...	52[e]	70[e]	...
Apr 1625. . . .	57	194	...	3.4	2.1	0.03	36.1	102	84	...
Jan 1655. . . .	52	320	...	6.2	5.2	0.3	24.4	114	144	63
Jan 1671. . . .	35	358	...	10.1	9.0	4.3	19.8	171	109	78
Dec 1672. . . .	26	267	...	10.2	10.8	2.0	22.4	167	86	14
Feb 1678. . . .	24	257[f]	...	10.7	9.8	0.4	21.6	198	58	0.4
Jy 1679	19	99[f]	...	5.2	5.3	1.0	16.8	...	...	...
Aug 1680. . . .	21	231	...	11.0	10.9	4.2	16.0	185	46	...
Dec 1682. . . .	27	286	...	10.6	10.6	3.3	17.0	201	79	6
Nov 1688. . . .	30	252	...	8.4	8.4	0.4	13.4	199	20	13
Aug 1718. . . .	27	287	...	10.6	11.0	0.5	15.9	202	70	15
Aug 1721. . . .	30	370	...	12.3	11.8	6.2	16.0	289	53	28
Dec 1746. . . .	26	288	...	11.0	10.8	0.2	32.2	231	45	12
Dec 1754. . . .	32	356	...	11.1	9.8	4.1	36.9	185	125	16
Dec 1763. . . .	45	554	...	12.3	11.2	2.5	29.6	311	144	109
Nov 1777. . . .	33	277	...	11.4	12.1	0.3	24.8	300	29	48
Jan 1786. . . .	35	409	...	11.6	12.1	4.0	...	289	48	72
Jan 1791. . . .	38	433	...	11.3	11.2	3.9	16.0	...	...	71
Feb 1807. . . .	48	537	...	11.2	11.8	3.1	17.5	...	0	...
Je 1811	33	360	...	10.9	10.6	5.1	15.2	338	0	22
Je 1812	41	384	...	9.4	9.6	1.9	16.0	...	0	...
Dec 1816. . . .	39	358	...	9.2	8.8	1.4	24.0	358	0	0
Oct 1821. . . .	28[a]	...	3,481	...	...	...	...	...	...	...
Feb 1829. . . .	31	...	3,297	...	...	...	...	...	...	...
Map, 1851? . .				10.6						

[a] Excludes all fields described as in poor condition. [b] Area listed in suertes. [c] Includes about 11 acres in Axomulco. [d] Includes second and third ratoons. [e] Tareas rather than acres; includes second and third ratoons in the case of ratoon cane. [f] Low because zafra just finished and planting not begun.

Table 12. Production and Yields of Sugar at Tlaltenango[a]

Year	Sugar Made, Lbs.	No. of Tareas	Lbs. per Tarea	Lbs. per Acre	Cartloads per Tarea
1555[b] . . .	25,683	25	1,027	935	...
1556[b] . . .	11,474	12	956	870	...
1557[b] . . .	27,432	22	1,247	1,370	...
1584[c] . . .	68,945	71	971	884	...
1587 . . .		112	1,084	986	25.0
1591 . . .	120,905	113	1,070	974	23.5
1592 . . .	113,424	106	1,070	974	22.4
1593 . . .	95,838	114	841	765	22.4
1594[d] . . .	90,992	106	858	781	22.0
1595[d] . . .	52,422	44	1,191	1,084	21.3
1596 . . .	120,978	107	1,131	1,029	20.6
1597 . . .	100,157	101	992	902	21.7
1598 . . .	53,738	62	867	789	20.7
1599 . . .	95,312	87	1,096	997	...
1600 . . .	93,368	86	1,074	977	21.4
1601 . . .	78,914	66	1,196	1,088	22.0
1602 . . .	91,872	83	1,107	1,209	...
1603 . . .	45,625	47	971		...
1604 . . .	50,126	43	1,166		21.1
1605[e] . . .	52,788	46	1,148		21.1
1606 . . .	38,676	49	789	718	...
1607 . . .	38,770	52	746	678	...
1608 . . .	45,609	62	736	669	...
1609[f] . . .	28,470	46	619	563	...
1610[f] . . .	15,773	39	404	368	...
1611 . . .	13,239	28	473	430	...
1612 . . .	2,684	6	447	407	...
1613[g] . . .	522	1	522	475	...
1614[h] . . .	12,397	21	590	537	...
1615 . . .	38,806	46	800	728	...
1616 . . .	12,996	18	722	657	...
1617 . . .	95,885	81	1,184	1,077	...
1619 . . .	67,399	93	725	660	...
1620 . . .	44,821	48	934	850	...
1621 . . .	105,850	95	1,115	1,014	...
1623 . . .	45,406	53	857	780	...
1624 . .	83,139	87	956	870	...

[a]Data are incomplete for 1618 and 1622. [b]Data from Axomulco records, L282 E4. [c]For sugar cane from other growers, used average weight of espumas to calculate weight of loaves of respumas. [d]To 9 July only. [e]After this date used muleloads rather than cartloads. [f]Includes October to December 1609. [g]Used average weight of loaf of 1612, 11.6 pounds. [h]Used average weight of loaf of 1615, 11.0 pounds.

Table 13. Prices and Kinds of Sugar Sold, 1779-81

Kind of Sugar	Pilón	Pedacería	Polvo
Blanco	15-17	14 1/2	...
Entreverada			
Superior or Blanco .	13 1/2-14	...	...
Entreverada	13-13 1/4	...	...
Entreverada Prieta .	12-12 3/4	...	...
Prieta	11-12 1/2	9 1/2-10 1/2	...
Zacate	...	...	8-8 1/2

Table 14. Eighteenth- and Nineteenth-Century Yields of Sugar

Year	Acres Cut	Sugar Produced, Lbs.	Lbs. per Acre
1768			
Jan-Mar	95	116,841	1,226
Sep-Oct	28	19,713	702
Total.	123	136,554	1,107
1769			
Jan-Sep	74	368,655	5,009
Oct-Nov	31	76,648	2,513
Dec	43	101,455	2,338
Total.	148	546,758	3,707
1770			
Jan	46	85,930	1,860
Mar-Apr	65	101,131	1,561
1779	156	147,813	947
1780	193	271,650	1,408
1782			
Jan-Apr	91	193,952	2,131
Summer.	42	77,173	1,837
Oct-Dec	80	79,574	995
1785			
Jan-Apr	53	77,652	1,465
Summer.	16	24,095	1,547
Oct-Dec	65	93,226	1,434
1795	244	434,564	1,781
1811	218	479,382	2,199
1848[a]	161	452,700	2,829
1851[a]	209	585,000	2,799
1851	209	633,038	3,029

[a]Predicted.

Table 15. Sizes of Kettles, 1682 and 1688

Kettle	Andanas	Depth, Ins.	Radius, Ins.	Approximate Volume, Gals.	1721: Weight, Lbs.
Caldera de					
Recibir	3	66	31	460	950
Contrarecibir	3	74	31	600	725
En medio.	3	58	31	380	850
Contramelar	3	66	31	460	900
Melar	3	74	31	560	800
Perol.	2	49	31	340	n.a.
Tacha (bronze)	0	22	16	32	125
Tacha (copper)	1	17	14	35	137
Tacha (copper)	1	17	17	45	350
Resfriadero (copper) . .	1?	28	20	80	n.a.
Two pailas (Fondos) for cachazas and claros. . .	n.a.	n.a.	n.a.	n.a.	n.a.

Table 16. Seasonal and Interannual Variations in Sugar and Molasses Production, 1610-25

Year	Lbs. of Sugar per Cart	Lbs. of Molasses per Cart	Total, Cols. 1 & 2	Lbs. of Miel de Furo per Cart	Total, Cols. 3 & 4	Estimated Values, Col. 5	Ratio of Sugar to Molasses, Lbs.	Ratio of Sugar to Miel de Furos, Lbs.	No. of Loaves per Cart
1610-11[a]	0	62	62	...	62	...	...	...	0.0
1611.	0	55	55	...	55	...	...	...	0.0
1612[b]	0	37	37	...	37	...	...	...	0.0
1616-17	42	10	52	15	67	...	.23	.35	3.7
Dry, 1 Jan-13 Apr..	...	7	...	...	...	...	...	...	...
Wet, 11 Jy-19 Aug. .	...	30	...	...	...	...	...	...	...
1617-18	33	21	54	39	93	...	.63	1.18	3.2
Dry, 1 Nov-1 Apr . .	...	16	...	...	...	...	...	...	...
1619.	31	13	44	19	63	61	.41	.61	2.8
Dry, 22 Jan-15 May .	35	9	43	...	...	...	...	...	...
Wet, 1 Jy-30 Aug . .	26	33	49	...	...	...	...	...	...
1620.	40	9	49	14	53	...	.22	.35	3.7
Dry, 4 Feb-7 May . .	42	8	50	...	...	...	...	...	...
Wet.	35	12	47	...	...	...	...	...	4.3
1621.	47	3	50	4	54	...	.06	.08	...
Dry, 4 Jan-30 Apr. .	50	6	56	...	...	...	...	...	...
Wet, 5 Jy-17 Aug . .	42	2	44	...	...	...	...	...	4.4
1622.	47	8	55	12	67	...	.17	.25	...
Dry, 26 Jan-14 Je . .	51	7	58	...	...	...	...	...	...
Wet, 5 Jy-17 Aug . .	41	9	50	...	...	...	...	...	...
1623.	45	4	49	32	81	95	.08	.71	3.9
1624.	44	13	57	27	84	86	.29	.61	4.2
1625, Dry, 10 Jan- 13 Mar.	52	17	69	...	...	101	.32	...	4.4

[a]Burned cane-molasses.   [b]Damaged or dry cane-molasses.

Table 17. Number of Loaves and Weight of Sugar Produced per Cartload of Cane, 1587-1625

Year	No. of Loaves per Cart	Lbs. of Sugar per Cart	Year	No. of Loaves per Cart	Lbs. of Sugar per Cart
1587. . .	4.1	43.1	1604. . .	4.3	50.7
1590. . .	3.5	n.a.	1605. . .	4.3	53.8
1591. . .	3.8	43.3	1606-7. .	4.2	47.9
1592. . .	4.2	47.5	1607-8. .	4.0	46.0
1593. . .	3.3	38.3	1616-17 .	3.7	42.0
1594. . .	3.4	38.4	1617-18 .	3.2	32.9
1596. . .	4.5	54.9	1619. . .	2.8	31.4
1597. . .	4.0	45.6	1620. . .	3.7	39.9
1598. . .	4.0	41.6	1621. . .	4.3	46.7
1600. . .	3.9	49.9	1622. . .	4.4	48.6
1601. . .	4.5	54.5	1624. . .	4.2	44.4
1602. . .	4.1	52.1	1625[a] . .	4.4	51.6
1603. . .	3.5	44.1			

[a]Winter only.

Table 18. Average Weights of Mill Equipment

Date of Inventory	Weight of Chumaceras, Lbs.	Weight of Camisas, Lbs.	No. of Calderas	Total Weight of Calderas, Lbs.	Average Weight per Caldera
1549. . .	50	...	7	...	...
1585. . .	...	...	7	...	...
1672. . .	...	...	8	...	...
1703. . .	70	...	...	...	...
1721. . .	...	...	8	4,725	591
1727. . .	60	...	...	...	...
1746. . .	...	...	10	10,700	1,070
1754. . .	...	1,317	9	11,325	1,258
1763. . .	85	...	...	...	...
1777. . .	...	1,000	10	14,350	1,435
1786. . .	70	1,375	10	14,300	1,430
1791. . .	60	1,400	10	13,100	1,310
1799. . .	60	1,550	10	14,550	1,455
1816. . .	...	...	11	...	...
1828[a] . .	...	1,900-2,875	...	...	...
1831[a] . .	125	2,650	...	...	...
1847. . .	...	2,775	12	...	1,375[b]
1884. .	...	...	14	17,500	1,250

[a]Data from annual account.
[b]Weight of only one kettle available.

Table 19. Numbers of Oxen[a]

Year	Total	Cart	Plow	Press
1549	123	...	...	...
1566-67	151[b]	...	...	...
1576	151	...	...	...
1585	307	270	37	0
1589	158	...	...	...
1600	50	...	...	...
1625	345	320	25	0
1655	469	352	109	8
1672	521	285	217	19
1679	657	407	236	14
1682	568	376	189	7
1688	551	333	203	21
1693	...	164	143	18
1718	...	112	143	...
1721	250[c]	...	...	...
1741	369?	369[d]	...	...
1746	192?	192	...	...
1754	115	35	80	0
1759	...	208	100	...
1763	299	199	100	0
1777	447	106	341	0
1791	520	80	440	0
1799	515	...	...	0
1807	658	107	551	0
1811	581	124	457	0
1818	358	...	...	0
1821	336	...	...	0
1829	246	...	...	0
1847	245[e]	...	...	0
1884	169	...	...	0

[a]For 1609-25, see Table 20.   [b]Cart and plow only.   [c]Includes novillos.   [d]Cañeros or tiros.
[e]Only surcadores and barbecheros.

Table 20. Mortality Rates of Oxen

Year	Months of Account	No. at Beginning	Died	Old, Culled	Lost	Bought	No. at End	Annual Mortality Rate[a]
1609	...	n.a.	n.a.	n.a.	n.a.	n.a.	226	...
1609-11 . . . 19.0		226	0	12	28	0	186	11%
1611-13 . . . 22.0		186	0	52[b]		0	134	15%
1613-15 . . . 29.5		134	109	0	0	358	383	33%
1615-17 . . . 20.5		383	n.a.	n.a.	n.a.	n.a.	482	n.a.
1618 . . . . 12.0		482	42[c]	0	0	0	440	8%
1619 . . . . 12.0		440	29	0	0	22[c]	433	6%
1620 . . . . 12.0		433	23	0	0	6	416	5%
1621 . . . . 12.0		416	18	32	0	0	366	12%
1622 . . . . 12.0		366	10	16	0	0	342	7%
1623 . . . . 12.0		342	13	0	14	100	415	7%
1624-25 . . 15.5		415	31	23	66	55	345	22%

[a]Average of all years: 13 per cent.
[b]Includes lost oxen.
[c]Estimated on the basis of available information.

Table 21. Numbers of Mules and Horses[a]

Year	Mules							Horses	
	Unspecified	Carga or Requa	Riding	Press	Carry Cane	Trapiche	Total	Males	Mares
1566	0	24	0	0	0	0	24	...	...
1585	0	37	0	0	0	0	37	...	...
1589	17	0	0	0	0	0	17	...	...
1655	0	0	18	4	0	0	22	65	56
1672	0	0	15	0	0	0	15	59	0
1679	0	0	19	0	0	0	19	52	12
1688	33	0	0	...	...	...	33	58	5
1718	342	...	...	...	...	...	342	...	...
1721	0	16	58	0	51	310	435	39	92
1741	14	16	67	0	97	213	407	0	0
1754	9	63	20	0	0	0	93	8	18
1763	0	70	49	0	169	0	288	25	22
1777	15	286	64	0	0	0	365	21	12
1786	12	277	77	0	...	0	366	56	34[b]
1791	28	244	94	0	...	0	366	47	28
1799	486	...	...	0	...	0	486	...	...
1807	20	445	131	0	...	0	596	46	21
1811	9	302	145	0	...	0	456	63	50[b]
1818	341	...	...	0	...	0	341	0	53
1829	291	...	...	0	...	0	291	21	59
1847	240	...	...	0	...	0	241	12	14
1884	181	...	...	0	...	0	181	7	0

[a]There was no mention of mules or horses in accounts for 1549.
[b]Includes colts.

Table 22. Carga Weights and Freight Costs

Year	Cargas to Capital	Weight in Arrobas	Weight per Carga, Lbs.	Cost per Carga, Pesos	Year	Cargas to Capital	Weight in Arrobas	Weight per Carga, Lbs.	Cost per Carga, Pesos
1591 . .	241	n.a.	238	1-4-0	1754 . .	22	n.a.	440	1-6-0
1595 . .	223	n.a.	250	1-4-6	1773 . .	986	19,606	497	2-2-0
1600 . .	391	n.a.	...	1-4-0	1782 . .	866	16,890	487	2-3-0
1644 . .	377	n.a.	...	1-0-0	1785 . .	827	15,454	467	2-3-0
1681 . .	336	n.a.	350	2-0-0	1787 . .	934	18,337	491	2-3-0
1707-8 .	832	12,301	370	2-0-0	1790 . .	731	13,973	478	2-4-0
1718 . .	403	6,700	415	n.a.	1795 . .	1,052	19,019	454	2-4-0
1719 . .	570	9,700	426	n.a.	1820 . .	158	n.a.	...	5-4-0
1720 . .	484	7,995	413	n.a.	1824 . .	781	n.a.	480	5-0-0

Table 23. Efficiency of Animal Use at Atlacomulco

Source	Year	Sugar Made, Tons	Oxen	Horses	Mules	Power
Dalby Thomas.	1690	50	8	6	0	Wind
Labat	1690's	200?				
Atlacomulco .	1688	150	550	58	31	Water
Belgrove. . .	1755	340	150	25	0	Wind
Atlacomulco .	1753-54	300	115	30	92	Water
Debien. . . .	ca. 1740-90	300	70	30	220	Animal
Long.	1774	Low	30	0	30	
Atlacomulco .	1770-95	400	450+	50	400	Water
Avalle. . . .	1788	300	16	0	105	Mules
Phillips. . .	1792-96	250	140	0	80	?

Table 24. Purchases of Imported Slaves

Date	Number	Source	Date	Number	Source
1579 . .	40	L247 E10	1615 . .	5	L269 E12
1598 . .	19	L263 E4	1619 . .	9	L269 E16
1599 . .	10	L263 E14	1620 . .	8	L269 E21
1600 . .	18	L263 E15	1623 . .	1	L269 E34
1614 . .	14	L269 E12			

Table 25. Caste and Place of Birth of Slaves

Race	Male	Female	Total
Negroes			
Born in Africa	263	75	338
Creoles	45	28	73
Total			411
Mulattoes	81	58	139
Cocho	19	18	37
Blanco	6	5	11
Prieto	5	3	8
Amulatado	1	0	1
Alobado	2	0	2
Total			198
Other	11	5	16
Caste and place of birth not identified	33	2	35
Creoles, caste not identified	112	117	229
Grand total			889

Table 26. Tribal Origins of African-Born Slaves

Country and Tribe	Male	Female	Total
Senegambia			
Biafara	35	16	51
Mandingo	20	3	23
Zape	16	8	24
Jalofa.	12	7	19
Banol	8	4	12
Berbesi	7	4	11
Nalu	4	3	7
Biojo	2	1	3
Cabo Verde	2	0	2
Cazanga	2	0	2
Jojo (Xoxo)	1	1	2
Balanta	1	0	1
Total			157
Ghana-Nigeria-Slave Coast			
Bran	28	10	38
Arda and Arara	9	3	12
Terra Nova	4	0	4
São Tomé	3	0	3
Carabali	2	0	2
Total			59
Congo and Angola			
Angola	37	13	50
Congo	10	0	10
Manicongo	8	1	9
Mocanga	2	0	2
Malemba	1	0	1
Matamba	1	0	1
Xigo	1	0	1
Total			74
East Coast			
Cibalo	0	1	1
Cafre	1	0	1
Zamuco	1	0	1
Mozambique	4	0	4
Total			7
Other	27	0	27
Grand total			324

Table 27. Maximum Annual Salaries Received by Skilled and Supervisory Personnel

Year	Mayordomo	Priest	Doctor	Purgador	Mandador	Cañaverero	Carretero	Sugarmaster
1580's. . . .	...	331	200	200	250	250	200	600
1590's. . . .	744	331	200	200	350	400	250	600
1600's. . . .	744	331	350	200	400	370	301	1,000
1610's. . . .	...	0	220	...	350	350	...	400
1620's. . . .	1,000	0	275	185	350	300	...	...
1640's. . . .	800	150[a]	100[a]	190	160	...	60[b]	None[c]
1681-82 . . .	700	...	...	...	160	...	...	None[c]
1702-4. . . .	500	...	...	...	...	...	...	...
1718-21 . . .	500	...	...	...	...	...	...	...
1779-1811 . .	500	...	60[a]	300	144	None	None	365[d]
1815.	1,000	...	...	300	144	...	...	...

[a]Called when needed; figure is approximate. [b]Paid to an Indian Capitán de carreteros. [c]Apparently this post was manned by Negro slaves. [d]Based on daily wage of $1.

Table 28. Wages and Contribution of Naborios

Class of Work	Average Annual Pay, Pesos		Average Daily Wage, Granos		Full-Time Equivalents		Actual Nos.	
	1581-93	1618-24	1581-93	1618-24	1581-93	1618-24	1581-93	1618-24
Carpenters	56	121	15.1	17.6	1	2	3	4
Smithy	97	0	8.2	...	5	0	8	0
Potters	66	86	4.4	7.8	4	4	8	5
Purgery	162	129	4.3	11.8	12	4	22	8
Laborers	75	...	7.7	...	3	...	6	...
Irrigation	47	47	4.1	8.6	4	2	8	3
Cart drivers	157	406	5.5	8.0	14	16	29	20
Oxherds	29	72	7.5	14.5	2	2	3	4
Ranch hands	171	78	8.6	17.8	7	1	16	4
Muleteers	113	...	11.9	...	3	...	6	...
Miscellaneous	52	73	5.0	10.0	3	2	7	4
Total	1,025	1,012			58	33	116	52
Average			7.5	12.0				

Table 29. Summary Concerning Naborío Labor

Year	Total Pay, Pesos	Total Days	Average Daily Wage, Granos	Full-Time Equivalent Workers	Actual Nos.	Ratio[a]
1581	959	16,900	5.5	56	114	1:2.8
1583	1,136	20,075	5.4	67	127	1:1.9
1584	1,250	20,575	5.8	69	98	1:1.4
1587[b]	1,579	24,837	6.1	83	143	1:1.7
1589[b]	1,116	15,200	7.1	51	109	1:2.1
1590[c]	1,023	16,512	6.0	55	109	1:2.0
1592	841	13,050	6.2	44	92	1:2.1
1593	n.a.	n.a.	n.a.	...	81	...
1602	1,130	12,112	9.0	40	62	1:1.6
1618[d]	942	7,575	11.9	25	60	1:2.4
1619[e]	970	7,828	11.9[f]	26	50	1:1.9
1620	1,068	8,612	11.9	29	55	1:1.9
1621	1,002	8,083	11.9	27	52	1:1.9
1622	926	7,467	11.9	25	60	1:2.4
1623	902	7,277	11.9	24	52	1:2.2
1624[g]	1,226	9,890	11.9	33	42	1:1.3

[a]Number of full-time equivalent workers over actual numbers.   [b]11 months.   [c]13 months.   [d]11 1/4 months.
[e]12 3/4 months.   [f]After 1618 it is necessary to assume that the average daily wage of 1618 applies.
[g]15 1/2 months.

134

Table 30. Length of Service of Naboríos

Class of Work	No. of Annual Accounts	Total No. of Persons	Average Length of Association, Years	Average Months Worked per Year
Purgery workers...	19	122	3.5	6.8
Cart drivers....	19	160	5.7	6.4
Muleteers.....	6	34	1.9	4.9
Mazatepec......	17	137	1.5	4.9
Irrigators.....	19	58	2.8	5.5
Potters.......	19	32	7.2	7.3

Table 31. Average Costs of All Kinds of Labor

Laborers	Average Days per Year	Average Cost per Year, Pesos	Average Cost per Day, Granos
Naboríos			
1583-1600[a]....	15,109	2,167	13.8
1621-24[b]....	8,254	1,150	13.4
Other Indians			
1583-1600[a]....	36,319	4,530	12.0
1621-24[b]....	5,167	1,346	25.0
All Indians			
1583-1600[a]....	51,428	6,697	12.5
1621-24[b]....	13,421	2,496	17.9
Negroes			
1581-1600[c]....	23,780	4,525	18.3
1621-24[b]....	20,400	4,636	21.8
Spaniards			
1581-1600[d]....	3,816	5,754	144.8
1621-24[b]....	2,449	2,246	88.0
All laborers			
1583-1600[e]....	79,795	17,055	20.5
1621-24[b]....	34,288	9,137	25.6
1780-1822[f]....	80,682	25,107	29.9

[a]14 years. [b]4 years. [c]15 years. [d]13 years. [e]12 years. [f]7 years.

Table 32. Productivity of Labor

Year	Man-Days	1,000 Lbs. of Sugar Made	Lbs. of Sugar per Man-Day	Lbs. of Sugar per Man-Year	Equivalent Full-Time Workers
1587...	91,661	122	1.3	399	306
1592...	75,350	126	1.7	504	251
1593...	74,093	110	1.5	447	247
1595...	80,073	107	1.3	399	267
1596...	71,945	116	1.6	483	240
1597...	69,935	100	1.4	429	233
1598...	77,626	67	0.9	261	259
1599...	78,900	100	1.3	381	263
1600...	90,226	107	1.2	357	300
1602...	69,395	106	1.5	459	231
1620...	36,979	45	1.2	363	123
1621...	37,415	106	2.8	849	125
1622...	33,206	88	2.7	795	111
1623...	33,049	45	1.4	411	110
1624...	33,481	83	2.5	744	112
1644...	36,612	69	1.9	561	122
1645...	45,251	70	1.6	465	151
1768...	45,011	177	3.7	1,110	150
1769...	59,872	584	9.3	2,790	200
1779...	69,046	233	3.4	1,020	230
1780...	63,169	312	4.9	1,470	211
1782...	72,995	409	5.7	1,710	243
1785...	76,618	447	5.8	1,740	255
1795...	78,062	506	6.5	1,950	260
1811...	82,486	551	6.7	2,010	275

Table 33. Total Labor and Productivity of Labor

Year	Negroes	Percentage of Total	Naboríos	Percentage of Total	Other Indians	Percentage of Total	All Indians, Percentage of Total	Spanish	Percentage of Total	Grand Total	Production of Sugar, 1,000 Lbs.	Lbs. of Sugar per Man-Day
1581...	23,400	31	16,900	23	31,320	42	65	2,887	4	74,507	...	...
1583...	24,900	31	20,075	25	30,373	38	63	4,050	5	79,398	...	...
1587...	24,000	26	24,837	27	38,112	42	69	4,712	5	91,661	122	1.33
1588...	22,200	27	15,560	19	39,684	48	67	4,600	6	82,044	...	...
1589...	22,200	26	16,587	19	42,827	50	69	4,675	5	86,289	...	...
1591...	21,600	...	16,228	...	35,078	...	...	...	...	...	138	...
1592...	21,600	28	13,050	17	37,793	50	67	3,507	5	75,350	126	1.68
1593...	21,000	28	12,755	17	37,127	50.	67	3,211	4	74,093	110	1.49
1594...	21,000	...	11,675	...	36,877	...	...	...	...	...	94	...
1595...	21,000	26	12,746	16	42,827	53	69	3,500	4	80,073	107	1.33
1596...	21,000	29	13,215	18	34,180	48	66	3,550	5	71,945	116	1.61
1597...	21,000	30	10,380	15	34,855	50	65	3,700	5	69,935	100	1.43
1598...	27,000	35	12,777	16	34,274	44	60	3,575	5	77,626	67	0.87
1599...	30,000	38	15,302	19	29,773	38	57	3,825	5	78,900	100	1.27
1600...	35,400	39	16,332	18	34,684	38	56	3,810	4	90,226	107	1.19
1602...	36,300	52	12,112	17	16,833	24	41	4,150	6	69,395	106	1.53
1620...	18,900	51	8,612	23	7,627	21	44	1,840	5	36,979	45	1.21
1621...	20,400	55	8,023	21	7,359	20	41	1,573	4	37,415	106	2.83
1622...	20,400	61	7,466	22	3,930	12	34	1,410	4	33,206	88	2.65
1623...	20,400	62	7,277	22	4,530	14	36	842	3	33,049	45	1.37
1624...	20,400	61	9,890	30	2,389	7	37	802	2	33,481	83	2.48
1644...	10,500	29	8,967	24	16,095	44	68	1,050	3	36,612	69	1.87
1645...	9,800	22	8,682	19	25,800	57	76	969	2	45,251	70	1.55
1779...	3,000	4	...	...	59,394	...	86	6,652	10	69,046	233	3.4
1780...	3,000	5	...	...	53,490	...	85	6,679	10	63,169	311	4.9
1782...	3,000	4	...	...	63,795	...	88	6,200	8	72,995	409	5.7
1785...	3,000	4	...	...	66,781	...	87	6,837	9	76,618	447	5.8
1795...	2,400	3	...	...	68,751	...	88	6,911	9	78,062	506	6.5

Glossary

Alcabala. An excise tax paid on goods sold or produced. Together with the tithe paid to the Church, it was the principal annual payment made by sugar plantations to the authorities.

Alcalde. The principal civil authority in provincial municipalities, with a wide range of civil powers.

Brazo. Equal to 2 varas, or 66 inches.

Clayed sugar. Sugar brought to a semirefined state by the use of water to dissolve and remove molasses coating the sugar crystals; the water was released slowly from moist clay placed on top of the loaf. It is no longer made, but in colonial times it was the most expensive kind of sugar made in Mexico.

Ingenio. A sugar mill powered by water and producing clayed sugar.

Legua (league). Apparently of variable length, but approximating 3 miles; Robelo (1908, p. 10) defines it as 5,000 varas or 4,190 meters.

Panocha. A hard brown sugar incorporating both sugar crystals and molasses, also called piloncillo. Still widely manufactured in Mexico and Central America, this was a principal product of colonial trapiches but was not made extensively in ingenios.

Tarea. A day's work, applied to an individual (see Appendix D) in plowing, or to an area of cane harvested by many workers in a day; as applied to harvesting in the sixteenth century, it was apparently equal to about 1.1 acre.

Temporal. Unirrigated land used for growing crops in the wet season.

Trapiche. A sugar mill powered by animals and generally producing non-clayed sugars; the milling equipment of any sugar mill.

Vara. Approximately 33 inches, or 0.838 meters (Robelo, 1908, p. 18).

Zafra. Harvest season on a sugar plantation.

Bibliography

Accounts of the Cortés Plantation

From		To		Source
	1541		1542	L277 E4 (Axomulco)
2 Aug	1553	2 Aug	1557	L282 E4 (Axomulco)
1 Jan	1581	31 Dec	1581	L257 E13
1 Jan	1582	31 Dec	1582	L257 E13
1 Jan	1583	31 Dec	1583	L257 E13
1 Jan	1584	31 Dec	1584	L257 E13
1 Jan	1585	31 Oct	1585	L257 E13; L262 E1
1 Nov	1585	31 Dec	1586	L262 E2 D6
1 Jan	1587	31 Dec	1587	L262 E2 D5
1 Jan	1588	31 Dec	1588	L262 E2 D7
1 Jan	1589	30 Nov	1589	L262 E2 D8
1 Dec	1589	31 Dec	1590	L262 E5 D2
1 Jan	1591	31 Dec	1591	L262 E5 D3
1 Jan	1592	31 Dec	1592	L262 E5 D4
1 Jan	1593	31 Dec	1593	L262 E5 D5
1 Jan	1594	31 Dec	1594	L262 E5 D6
1 Jan	1595	9 Jy	1595	L262 E5 D7
10 Jy	1595	31 Dec	1595	L263 E1
1 Jan	1596	31 Dec	1596	L263 E3 C1
1 Jan	1597	31 Dec	1597	L263 E5
1 Jan	1598	31 Dec	1598	L263 E4
1 Jan	1599	31 Dec	1599	L263 E14
1 Jan	1600	31 Dec	1600	L263 E12
1 Jan	1601	31 Dec	1601	L263 E13
1 Jan	1602	31 Dec	1602	L263 E8; E13; E14
5 Feb	1603	10 Apr	1606	L263 E12
11 Apr	1606	30 Sep	1607	L268 E1
23 Oct	1607	31 Dec	1608	L268 E3
1 Jan	1609	26 Sep	1609	L268 E3B
27 Sep	1609	1 May	1611	L268 E2
2 May	1611	21 Feb	1613	L268 E4
22 Feb	1613	15 Aug	1615	L269 E1
30 Oct	1616	31 Dec	1617	L269 E12
1 Jan	1618	31 Dec	1618	L269 E14
1 Jan	1619	31 Dec	1619	L269 E16
1 Jan	1620	31 Dec	1620	L269 E21
1 Jan	1621	31 Dec	1621	L269 E22
1 Jan	1622	31 Dec	1622	L269 E31
1 Jan	1623	31 Dec	1623	L269 E34
1 Jan	1624	16 Apr	1625	L269 E36
11 Jan	1644	30 Jan	1645	L257 E7 D1
30 Jan	1645	31 Jy	1645	L257 E7 D2
1 Aug	1645	31 Dec	1645	L257 E7 D3
1 Sep	1681	31 Aug	1682	L257 E15; L148 C1
25 Feb	1702	31 Dec	1704	L30 V56 E1
20 Aug	1718	20 Aug	1721	L30 V57 E1; E2; E3
1 Jan	1754	17 Dec	1754	L341 E3; E4
1 Jan	1768	31 Dec	1768	L341 E15
1 Jan	1769	31 Dec	1769	L341 E15
1 Jan	1770	28 Apr	1770	L341 E15

Bibliography of Documents and Printed Sources

THE citations of documents are abbreviated in form, a feature made possible by a nearly exclusive dependence on documents contained in the Hospital de Jesús collection of the National Archive of Mexico. As in other branches of the Archive, bundles of documents in the Hospital de Jesús collection are called legajos, which I indicate throughout this work by the letter *L*. The legajos are subdivided into expedientes (*E*), cuadernos (*C*), documentos (*D*), and, in the cases of some legajos with low numbers, into volúmenes (*V*); where a document has been appended to a legajo for some reason, it may be called an atado (*A*). In many expedientes in the Hospital de Jesús collection, the various documents are numbered separately rather than serially; for this reason, I have omitted page numbers in the citations. The letter *T* refers to volumes from the Tierras collection in the Archive, and I have included in this work information from two volumes of the small Padrones collection.

Until about forty years ago, the Hospital de Jesús (a block south of the Zócalo) housed the collection of Estate documents now known by its name. Information about the history of the Hospital is given by Alamán, who supervised its refurbishing in the early nineteenth century and took justifiable pride in the results (*Documentos* and *Disertaciones, passim*); he appears to have been thoroughly familiar with the records of the Hospital and the Estate. The site of the building he described is now occupied by a much larger Hospital de Jesús in a modern building. Charitable foundations such as the Hospital were common in New Spain; for example, Isabel de Ojeda, owner of Axomulco, appears to have started such a hospital (L27 V48 E3), but it may have been unsuccessful because of the incorporation of her plantation within Tlaltenango by Martín Cortés. Doubtless, the preservation of the records of the Estate, including those of the Hospital itself and the plantation, was greatly enhanced by their retention in the Hospital, for they contain a relatively large number of sixteenth- and seventeenth-century documents; the eighteenth century is fairly well represented, but there are relatively few documents dating from the period following disentailment of the Estate. In respect of the plantation and urban properties this lack of nineteenth-century documents is rather puzzling, because Alamán was insistent on the importance of keeping good records.

From		To		Source
1 Jan	1779	31 Dec	1779	L387 E35
1 Jan	1780	31 Dec	1780	L387 E35
1 Jan	1782	31 Dec	1782	L387 E36
1 Jan	1785	31 Dec	1785	L232 E2 D4
4 Jan	1795	2 Jan	1796	L378 E5
31 Dec	1810	8 Je	1811	L30 V55 E5
9 Je	1811	31 Dec	1811	L30 V55 E4
1 Jan	1815	31 Dec	1815	L30 V55 E9
3 Jan	1819	1 Jan	1820	L30 V54 E3
6 Jan	1822	4 Jan	1823	L30 V54 E4
6 Jan	1828	13 Dec	1828	L233 E7
1 Jan	1831	31 Dec	1831	L457
1 Jan	1832	31 Dec	1832	L457
4 Jan	1847	2 Jan	1848	L390 E3

Inventories of Tlaltenango and Atlacomulco

10 July 1549. México. Archivo General de la Nación, *Publicaciones*, Vol. 27. *Documentos Inéditos Relativos a Hernán Cortés . . .*, pp. 250–83. México, 1935. Contains inventories of all Cortés' possessions.

1 January 1567. L276 E77.
1 January 1576. L282 E2.
29 October 1585. L262 E2; L262 E2 D3.
6 December 1589. L262 E5 D1.
17 April 1606. L263 E15.
23 October 1607. L268 E3B.
28 September 1609. L268 E2.
2 May 1611. L268 E3.
21 February 1613. L269 E14 DE.
5 December 1618. L269 E14 DF.
16 April 1625. L38 V72 E18; L28 V50 E9; L269 E40.
21 May 1634. L38 V72 E18.
8 January 1655. L269 E42.
21 January 1671. L269 E44; L93 E2.
31 October 1672. L269 E45.
29 December 1672. L93 E14.
31 January 1678. L93 E2.
29 July 1679. L93 E2.
28 August 1680. L93 E7.
29 December 1682. L30 V53 E1.
3 November 1688. L30 V53 E2.
22 February 1693. L93 E1.
22 August 1718. L341 E1.
10 September 1721. T1965.
11 August 1741. T1965.
16 December 1746. T1965.
13 December 1765. L341 E5.
29 December 1763. L341 E10.
31 December 1767. L341 E2.
4 November 1777. L30 V53 E3.
30 January 1786. L30 V53 E5.
11 January 1791. L402 E27.
September 1799. L402 E38.
19 February 1807. L402 E22.
10 June 1811. L376 E10.
19 June 1812. L402 E16.
3 December 1816. L244 E34.
11 September 1818. L244 E29.
23 October 1821. L244 E30.
13 February 1829. L402 E33.
1 November 1847. L402 E35.
1 January 1885. L402 E37.

Bibliography of Printed Works

Aguirre Beltrán, Gonzalo. *La población negra de México, 1519–1810*. Mexico City, 1946.

Alamán, Lucas. *Disertaciones*, 4 vols., ed. by Rafael Aguayo Spencer. Mexico City, 1942.

——. *Documentos diversos*, 4 vols., ed. by Rafael Aguayo Spencer. Mexico City, 1945–47.

Antonil, André João [João Antonio Andreoni, S.J.]. *Cultura e opulencia do Brasil por suas drogas e minas*. São Paulo, 1922.

Archivo General de la Nación (México). *Documentos inéditos relativos a Hernán Cortés y su familia*. Publicaciones, No. 27. Mexico City, 1935.

Avalle. *Tableau comparatif des productions des colonies françaises aux Antilles, avec celles des colonies anglaises, espagnoles et hollandaises de l'année 1787 à 1788*. Paris, An VII.

Baker, John P. *An essay on the art of making muscovado sugar*. Jamaica, 1775.

Banco de México, S.A. *La industria azucarera de México*, 3 vols., Mexico City, 1952–55.

Barrett, Ward. "Caribbean sugar-production standards in the seventeenth and eighteenth centuries." In John Parker, ed., *Merchants and scholars*. Minneapolis, 1965.

Berthe, Jean-Pierre. "Xochimancas: Les travaux et les jours dans une hacienda sucrière de Nouvelle-Espagne au XVIe siècle." *Jahrbuch für Geschichte von Staat, Wirtschaft und Gesellschaft Lateinamerikas*, Vol. 3 (1966), pp. 88–117.

Borah, W. W. *Silk raising in colonial Mexico*. Ibero-Americana, No. 20 (1943).

——. *Early colonial trade and navigation between Mexico and Peru*. Ibero-Americana, No. 38 (1954).

—— & S. F. Cook. *Price trends of some basic commodities in central Mexico, 1531–1570*. Ibero-Americana, No. 40 (1958).

——. *The aboriginal population of central Mexico on the eve of the Spanish Conquest*. Ibero-Americana, No. 45 (1963).

Cadenhead, I. E. "Some mining operations of Cortés in Tehuantepec, 1538–1547." *The Americas*, Vol. 16 (1960), pp. 283–87.

Calderón de la Barca, Fanny. *Life in Mexico: The letters of Fanny Calderón de la Barca*, ed. by Howard T. & Marion Hall Fisher. Garden City, N.Y., 1966.

Chaunu, Huguette & Pierre. *Séville et l'Atlantique, 1504–1560*, 8 vols. Travaux et mémoires de l'institut des hautes études de l'Amérique latine, No. 4 (1955–59). Paris.

Chevalier, François, ed. *Instrucciones a los hermanos jesuitas administradores de haciendas*. Universidad Nacional Autónoma de México, Instituto de Historia. Publicaciones, Ser. I, No. 18. Mexico City, 1950.

——. *La formación de los grandes latifundios en México. Problemas agrícolas e industriales de México*, Vol. 8 (1956), No. 1.

Cline, Howard. "Documentos pictóricos de los indios mexicanos." *Boletín del Archivo General de la Nación* (México), Vol. 4 (1963), No. 1, pp. 5–56.

Conway, G. R. G., ed. *Postrera voluntad y testamento de Hernando Cortés, Marqués del Valle*. Mexico City, 1940.

Daubrée, Paul. *La question coloniale au point de vue industriel*. Paris, 1841.

Debien, G. *Une plantation de Saint-Domingue. La sucrerie Galbaud du Fort (1690–1802)*. L'Institut français d'archéologie orientale du Caire, Cairo, 1941.

Deerr, Noël. *The history of sugar*, 2 vols. London, 1949–50.

Diderot, Denis. *Encyclopédie, ou Dictionaire raisonné des sciences, des arts et des métiers. Recueil des Planches . . .*, Vol. 1. Paris, 1762.

Diez, Domingo. *Bosquejo histórico geográfico de Morelos*. Cuernavaca, Morelos, 1967.

Dutrône la Couture, Jacques François. *Précis sur le canne et sur les moyens d'en extraire le sel essentiel*. Paris, 1790.

Fernández de Recas, Guillermo. *Mayorgazgos de la Nueva España*. Mexico City, 1965.

Fries, Carl. *Geología del Estado de Morelos y de partes adyacentes de México y Guerrero . . .* Universidad Nacional Autónoma de México, Instituto de Geología, Boletín No. 60. Mexico City, 1960.

Gibson, Charles. *The Aztecs under Spanish rule; a history of the Indians of the Valley of Mexico, 1519–1810.* Stanford, Calif.: Stanford University Press, 1964.

Gómez de Cervantes, Gonzalo. *La vida económica y social de Nueva España,* ed. by Alberto María Carreño. *Biblioteca Histórica Mexicana,* Vol. 19. Mexico City, 1944.

Haven, Gilbert. *Our next-door neighbor: A winter in Mexico.* New York, 1875.

Hernández Sánchez-Barba, Mario, ed. *Hernán Cortés: Cartas y documentos.* Mexico City, 1963.

Labat, Jean Baptiste. *Nouveau voyage aux îles de l'Amérique.* Paris, 1722, 6 vols.

Lamb, Robert B. *The mule in Southern agriculture.* University of California, *Publications in Geography,* Vol. 15 (1963).

Lewis, Oscar. *Life in a Mexican village: Tepoztlán [Morelos] restudied.* Urbana, Ill., 1963.

Ligon, Richard. *A true and exact history of the island of Barbados.* London, 1657.

Lippmann, Edmund O. von. *Geschichte des Zuckers . . .* Leipzig, 1890.

Long, Edward. *The history of Jamaica,* 3 vols. London, 1774.

Luis (Noriega). *Breves apuntes sobre el cultivo de la caña de azúcar en el Estado de Morelos.* Tepoztlán, Morelos, 1882.

Mauro, Frédéric. *Le Portugal et l'Atlantique au XVIIᵉ siècle (1570–1670).* Paris, 1960.

Mazari, Manuel. "Un antiguo padrón itinerario del Estado de Morelos." Sociedad Científica "Antonio Alzate," *Memorias,* Vol. 48 (1927), pp. 149–70.

Morelos, Gobernador. *Memoria presentada al H. Congreso del Estado.* Cuernavaca, Morelos, 1873.

———. *Memoria sobre el estado de la administración pública de Morelos.* Cuernavaca, Morelos, 1882.

Morelos, Secretaria general de gobierno, Sección de Hacienda. *Informe general del ramo por todas las operaciones practicadas durante el año de 1872 . . .* Cuernavaca, Morelos, 1873.

Phillips, U. B. "A Jamaica slave plantation." *American Historical Review,* Vol. 19 (1914), pp. 543–47.

Reti, Ladislao. "The Codex of Juanelo Turriano (1500–1585)." *Technology and Culture,* Vol. 8 (1967), pp. 53–66.

Robelo, Cecilio A. *Revistas descriptivas del Estado de Morelos.* Cuernavaca, Morelos, 1885.

———. *Diccionario de pesas y medidas Mexicanas.* Cuernavaca, Morelos, 1908.

Ruiz de Velasco, Felipe. *Historia y evoluciones del cultivo de la caña y de la industria azucarera en México . . .* Mexico City, 1937.

———. "Bosques y manantiales del Estado de Morelos." Sociedad Científica "Antonio Alzate." *Memorias,* Vol. 44 (1925), pp. 113–63.

Sandoval, Fernando B. *La industria del azúcar en Nueva España.* Universidad Nacional Autónoma de México, Instituto de Historia, *Publicaciones,* Ser. 1, No. 21. Mexico City, 1951.

Suárez de Peralta, Juan. *La conjuración de Martín Cortés . . .,* ed. by Agustín Yáñez. Universidad Nacional Autónoma de México, Biblioteca del Estudiante Universitario, No. 53. Mexico City, 1945.

Thomas, Dalby. *An historical account of the rise and growth of the West India colonies.* London, 1690.

Thornthwaite Associates. *Average climatic water balance data of the continents. Part VI, North America.* Laboratory of Climatology. *Publications in Climatology,* Vol. 17, No. 2. Centerton, N.J., 1964.

Velasco, Alfonso Luis. *Geografía y estadística del Estado de Morelos. Geografía y estadística de la república Mexicana,* Vol. 7. Mexico City, 1890.

Ward, Henry. *Mexico in 1827,* 2 vols. London, 1828.

Zavala, Silvio, & María Castelo. *Fuentes para la historia del trabajo en Nueva España,* 8 vols. Mexico City, 1939–48.

Maps and Aerial Photographs

It was not until the mid-eighteenth century that the idea of mapping plantations became fairly common in Morelos. The first map of Atlacomulco was made by the surveyor Alarcón in 1760. It contains general information about sites of neighboring pueblos and other properties, shows in detail only the area in cane, and gives no other indication of the boundaries of the plantation. Unfortunately, in spite of the surveyor's care, the shape of the area in cane must be grossly inaccurate; part of the trouble may be due to the fact that his indicated north is about 40 degrees east of the north shown on both the 1850 map and the aerial photographs. The major result was greatly to reduce the area of the figure, but by altering the scale of his map, it is possible to make segments of his perimeter fit fairly well to the 1850 map. The map is in L27 V48 E3, a document concerning the long-lived Ojeda suit.

The first accurate map of the plantation was ordered by Lucas Alamán about 1850, after completion of another that he found unsatisfactory. Both are uncatalogued in the Hospital de Jesús collection.

Maps from the collection at the Observatory in Mexico City are referred to by their catalogue numbers. J. L. Tamayo has compiled a series of maps of Mexican states, published in Mexico City. The map of Morelos appears to date from the late 1950's, and is at a scale of 1:250,000.

Aerial photographs were made in 1959 by the Compañía Mexicana Aerofoto, México, D.F.

INDEX

Index

Chichinautzin volcanics: nature and distribution of, 26; relation to alluvium, 28
Chiconcuac: trapiche, 57
Chocolate: ration of, 95
Cholera: epidemic of *1850*, 84
Cinnamon: ration of, 96
Clarification: of cane juice, 56–57
Clayed sugar: distinguished from non-clayed, 50
Claying: description, 59; kinds of molasses produced by, 61–62
Clothing: ration of, 96–97
Coa: uses and numbers of, 45
Coffee: production planned, 52
Coins: fluctuations in value, 21
Colegios: sources of cash, 6
Colonies: and sugar, 3
Columbus, Christopher: and Hispaniola, 3
Concrete: sugar, 50
Confectioners: of Mexico City, 23
Congo, 79
Conserves: produced at mill, 60
Controller: of Estate, duties, 15; and libranzas, 22
Convents: as sources of cash, 6
Copper: uses and costs, 73
Corn. *See* Maize
Cortés, Fernando: bankruptcy of, 13, 14
Cortés, Hernán: economic activities of, 8–13; and Axomulco, 38; irrigation system named for, 42–44; importation of slaves by, 78; establishment of plantation, 103
Cortés, María: dowry of, 14
Cortés, Martín: accused of treason, 13–14; economic activities of, 14; and Axomulco, 38, 39
Cortés, Pedro: and censos, 11–12; and Governor, 14
Cortés Estate. *See* Estado del Valle
Costs: of livestock, 70–72; of iron, steel, copper, 73; of Spanish labor, 78; of rations, 93–97 *passim*; of slaves, 97–98; of labor, and viceregal edicts, 99–100; of labor, in relation to all costs, 100–1. *See also* Freight costs
Coyoacán: claims of Cortés in, 9; and silkworm raising, 11; College and Convent of, 13
Criolla cane, 47. *See also* Cane
Crystallization: of sugar, 58
Cuautla: sugar mills in, 5; mines, 87
Cuba, 3–4
Cuernavaca: population changes, 10; ejido of, 36; manufacture of aguardiente, 61; source of Indian labor, 87; maize tributes from, 95
Cuernavaca formation: nature and distribution, 25–26; and site of Axomulco, 28
Cuernavaca-Cuautla basin: importance of, 4
Cycles: of sugar production, 5

Dams: role in water supply, 41
Debt peonage: and plantation store, 23–24; effectiveness of, 91–92
Denmark: and Caribbean sugar industry, 3
Diseases: of cane, in Morelos and Puebla, 45–46
Dispenser: duties of, 75; records of, 121
Diversion dams: role in water supply, 41
Doctors: duties of, 76
Doubleuse: use, 53
Drafts. *See* Libranzas
Drainage: in Cortesian irrigation system, 44
Drying: of sugarloaves, 60
Dung. *See* Soil fertility

Egurén, Andrés de, 34
Ejidos: disputes concerning, 35, 36
England: and early Caribbean sugar industry, 3. *See also* British West Indies
Entreverada: sugar, description of, 60–61

Epidemics: effects of on population size, 9–11, 104; among oxen, 67; effects of on Negro slaves and Indians, 82
Espumas: manufacture, 60
Estado del Valle (Cortés Estate): sugar plantations in, 7; extent and location, 8, 9; origins and entailment, 9; composition of income, 9–12, 103; urban properties, 12; changes in income, 12–13; sequestration in *1567*, 13; disentailment, 13, 17; legal jurisdiction, 14–15; duties of officials, 14–17; income confiscated in *1831*, 17
Evapotranspiration. *See* Water need

Fallowing: in Morelos, 45
Felipe, Diego, 77
Fernández de Mata, Juan, 77
Fertilizers. *See* Soil fertility
Field size: range of, 48
Figueroa, Gregorio de, 76
Filters: for cane juice, 57
Financing plantation operations: role of merchants and religious bodies, 6; sources of cash, 21–24
Fire: role of cañaverero, 76
Firewood: supply, 66; and consumption, 72–73; and springs, 73; sold by employees, 76, 77
Fish: ration of, 97
Flota: and cash supply, 21
Flour mill: at Tlaltenango, 42
Forms: description and use, 58, 59; size in Morelos, 60; made by naboríos, 89
France: and early Caribbean sugar industry, 3. *See also* French West Indies
Freight costs: paid in Mexico City, 23; changes in carga and, 71
French West Indies: arrangement of kettles, 56; efficiency of animal use, 72
Frost: and lessees, 20; hazard, 25; use of cane damaged by, 61
Fuel. *See* Bagasse; Firewood

Gañanes: labor inputs, 102
García, Baltasar, 77
García, Simón, 77
Gente de razón: residence at Tlaltenango, 38–39; wage and livestock preferences, 70; and transport of sugar, 72
Geology: of Morelos, 25–26
Ghana, 79
Gold: mining of, by Estate, 11
Governors of Estate: duties and powers of, 14–15; replaced by Apoderado General, 16; and plantation, 24; and discipline of labor, 84–86
Grazing: conflict with temporal, 35
Guadalupe: spring of, 39
Guarapa: definition, 57
Guatemala, 86
Guatetelco: lake, 34
Güell, Antonio, 16
Guimac, Rancho de: geology, soils, use, 26; purchase, 29, 32; exchange, 30, 36

Habas: rations, 97
Haciendas: and pueblos, 5; nature of, 6, 7
Haciendas Marquesanas: source of livestock, 11, 34; sale of, 17. *See also* Tehuantepec
Handbooks: of plantation operations, 7; and expectable yields, 40
Harriero. *See* Muleteer
Harvesting: and age of cane, 47; description, 47–48; labor inputs, 102
Havana: transport costs in, 4
Hawaii: as sugar colony, 3; irrigation in, 44
Hides: production, 65
Hispaniola, 3
Hoe: uses and numbers of, 44–45
Horses: uses and numbers of, 69–70; values of, 70–71; maize rations, 95. *See also* Livestock
Hospital de Jesús: properties in Mexico City, 12; liability of Estate

for, 13; in nineteenth century, 16; loss of income, 17; patronage of, 17; loans to mill, 21; rented store, 23
Huerta, La: purchase of, 29, 30–31; water rights of, 30
Hydrometer: use of in striking, 57

Independence: and Estate, 13, 16; and supply of cash, 21; and mill, 47, 66; and labor, 87–88
Indian labor: ownership of tools, 45; replacement of by oxen, 67; wage and livestock preferences, 70; importance of, 78; characteristics of, 86–89; supervision of, 88; payment of, 88–89; and debt peonage, 91–92; and viceregal edicts, 92; rations of, 93–97 passim. See also Naboríos
Indian labor supply: affected by viceregal edicts, 14, 99–100; guaranteed to lessees, 20; effects of Independence on, 24; related to planting schedule, 46; procured by mandadores, 76; from Cuernavaca jail, 86
Indian labor uses: to make forms, 59; to produce supplies, 65–66; in transport, 66; to supply firewood, 72
Indian principales: and labor supply, 87–89
Indian slaves: numbers of, 86
Indigo: production planned, 52
Ingenio: distinguished from trapiche, 54
Inventories: characteristics of, 19
Iron: substituted for wood, 63–64; use and costs, 73
Irrigation: description, 40–44; water need and costs, 42; water available for, 43; advantages of, 44; and plowing, 45; naboríos and, 90; labor inputs, 102
Irrigation network: of Cortés plantation, 28, 39
Irrigation routine: in Puebla, Oaxaca, and Morelos, 42–44; in Hawaii, 44
Istayuca (estancia): and water rights, 38
Istayuca, Santa María de: importance of springs near, 26, 28

Jalapa, 11
Jamaica: as sugar producer, 3; compared with Morelos, 7. See also British West Indies
Jamaica train: not used in Morelos, 56; fuel savings in, 73; disadvantages of, 105
Jerga: of Michoacán, in filters, 57
Jesuits: and colonial sugar industries, 3
Jiutepec: as owner and lessor of land, 28, 29, 31, 32, 37
Junta: of the Estate, duties, 16

Kettles: description, 55–56; construction, 62–63; changes in, 62–63, 73; use of copper and repairs, 73; importance of changes in weight, 105

Labor: and the planting schedule, 46; inputs, 74; treatment of, 84–86; seasonal aspects, 90–91; rations of, 93–97, 100; hours worked, 99; total costs and inputs, 99–100; changes in productivity, 101–2. See also Gente de razón; Indian labor; Mestizos; Mulatto slaves; Naboríos; Negro slaves
Labor turnover: of Spaniards, 74–75; of overseers and cañavereros, 76–77; of naboríos, 90–91
Labrador: duties of, 76; post eliminated, 77; naborío assistant to, 89–90
Land: leasing, 28; characteristics and importance of leases, 28–29; purchases, 29–36
Land disputes: difficult to settle, 32; causes, 36; after Independence, 37–38; role of race in, 37–38. See also Boundary disputes
Landforms: of Morelos, 25–26
Lawyer: of Estate, duties, 15
Leases: characteristics, 18–21; profitability, 21
Lessees: characteristics of, 6, 20–21, 110–13; and bondsmen, 18–19; and boundaries, 35; and improvements, 62; and livestock, 66; and resident labor, 91
Libranzas: nature and importance of, 21–23; and cash, 24
Livestock: in Hispaniola, 3; from Estate ranches, 11; and Atlacomulco pastures, 36; uses of, 66–72; costs of, 70–72; efficiency of use, 72; tended by naboríos and Negro slaves, 90

Loans: to muleteers, 72
Lomelín, Leonardo: importer of slaves, 78, 81
López del Castillo, Francisco, 34
López Morgado, Juan, 34
Lye: production and uses, 65

Machete: use, 47–48
Maestro de azúcar. See Sugarmaster
Madeira: as sugar producer, 3, 4; compared with Morelos, 7
Maintenance: responsibility for, 20
Maize: supply to lessees, 20; tributes of supplied to plantation, 65; and labor supply, 88; costs and rations of, 95; production at plantation, 95; sold by Estate, 95
Malemba, 79
Mancuernas: introduced, 63
Mandingo, 79
Manure. See Soil fertility
Mapping: of plantation, 35
Maquila: and priests, 76
Marqués, Damián, 76
Marquesado del Valle. See Estado del Valle
Martín, Diego, 77
Martín, Gaspar, 76
Mascabado. See Muscovado
Matamba, 79
Mateos, Cristóbal: suit with Jiutepec, 31; and pastures, 36; and water rights, 38–39
Mayordomo: of Estate, change of title to Governor, 14
Mayordomos of Atlacomulco: duties and income, 75; and discipline of labor, 84–86; food allowance, 96; improvement in skills, 104; records of, 118–21
Mazatepec, Estancia de: early importance, 11; location, description, and use, 34; end of dependence on, 39; and livestock supply, 66; and mules, 69; and horses, 70; labor at, 90, 91
Meat supply of Cuernavaca: and Mazatepec, 34; and local pastures, 36; and Atlacomulco, 65
Medical care: of slaves, 76
Melado: definition, 56
Meladura: description and processing, 57–58
Merchants: and cash supply, 6; and hacendados, 20, 23
Mesa, Manuel de, 84
Mestizos: importance as labor, 5, 104; and planting dates, 46
Metals. See Copper; Iron; Steel
Mexico: colonial sugar production, 4
Mexico City: as focus, 7, 25; and claims of Cortés, 9; Estate properties in, 13; sales of sugar in, 23; as source of slaves, 78
Miacatlán: hacienda, 34
Michoacán: as colonial sugar producer, 4; as source of filters, 57; as source of labor, 88
Mill: plan of, 50–53; construction of, 53–54; improvements in, 62–64; Indian and Negro labor in, 92. See also Trapiche
Millán de Velasco, Marcos, 76
Milling: rates reviewed, 54–55; labor inputs, 102
Mines: of Estate, 11; and labor, 87
Mixteca, 85
Mocanga, 79
Molasses: sale, 23–24, 61–62; and alcoholic beverages, 24; lack of data, 40; production, 48–49, 61–62; kinds and uses, 60–62; prohibition of sale of, 61; production records, 75
Molina, Cristóbal de: Governor, 14
Monteleone and Terranova, Duke of: heir of Hernán Cortés, 16–17, 103
Morelia, 4
Morelos: current production in, 4; comparisons, 7; and claims of Cortés, 9
Mortality rates: of oxen, 66–67
Mortgages: importance in Morelos, 5
Movellán y Lamadrid, Joseph de, 32
Mulatto slaves: definition and classifications of, 79; importance of, 79, 98; and flight, 85, 104; values of, 98

and sugarmaster, 77; in pottery, 77–78; changes in importance of, 78, 104; rations of, 93–97 *passim*

Springs: distribution of in Morelos, 25; importance of, 38–39; of Guadalupe, 39; and diversion dams, 41; affected by forest cutting, 73

Steam: use of in processing cane and juice, 63

Steel: use and price of, 73

Storage dams: role in water supply, 41

Striking: of cooked juice, 57–58

Sugar: and military and imperial strategy, 3; persistence and dominance, 5; importance of and experiments in tropics, 6–7; technical changes in production, 7; quality of, 23, 61; quotas in Morelos, 48; kinds of, 49, 50, 60–61; and molasses production, 62; and firewood, 72; production and shipment, *1765–95*, 102. *See also* Beet sugar

Sugar cane. *See* Cane

Sugar prices: and California gold rush, 11, 63; in seventeenth and eighteenth centuries, 21; and kinds produced, 60–61; and values of slaves, 82; and costs of corn and mutton, 95; trend, 103

Sugar sales: rent of store, 20; and cash supply, 21; and store administration, 22–23; by action, 23

Sugarloaves: weights of, 48–49, 71; drying and packing of, 60

Sugarmaster: duties and characteristics of, 77; Indians and Negroes as, 104

Sultepec: mines, 11

Supplies: and self-sufficiency of mills, 65–66

Surinam, 3

Swamps: and soils and terrain, 41–42

Tachas: use of, 57

Tahitian cane: experiments with, 46

Tallow, 54

Tannery: of Marqués, 11

Tarea: and plowing, 45; and milling rate, 54; changes in, 105

Tareas: labor inputs, 102

Taxco: Estate mines in, 11; and livestock supply, 34; and labor, 87

Teaches: use of, 57

Tehuantepec: Isthmus of, and Estate, 9; gold mines and livestock, 11; sale of properties, 17; production of mules, 69, 70. *See also* Haciendas Marquesanas

Tejalpa: pueblo, 36–37; and labor, 87, 88

Temixco: ingenio of and Mazatepec, 34; as neighbor, 35; population, 91

Temperature: regime in Morelos, 40–41

Temporal: conflict with grazing, 35

Tenapantles: in irrigation, 43

Tendidas: in irrigation, 43–44

Tepostlán: villa, and subject, 35; and labor, 87

Terranova, Duke of. *See* Monteleone and Terranova, Duke of

Tesontepec, 29

Tetela (pueblo), 28

Tintepec (pueblo), 35

Tlacomulco: purchase of, 29–30, 39. *See also* Atlacomulco

Tlalcoalpanque (trapiche), 69

Tlalhuapan: purchase of, 29, 31–32, 37

Tlaltenango: as site, 11, 25, 26–28, 103; obraje at, 11, 86; rental price in *1570*, 12; sugar sent to Peru, 12; entailed, 14; flour mill at, 42; free resident workers of, 91

Tlaltizapán (ranch): production of horses, 11; production of mules, 69–70

Tobacco: costs and rations of, 97

Toluca: claimed by Cortés, 9; and maize supply, 65, 95

Tools. *See* Coa; Hoe; Machete; Plows

Transport: effects on sugar, 61; by Indians, 66; by mules, 69; labor inputs, 102. *See also* Freight costs

Trapiche: construction of, 53–54; distinguished from ingenio, 54; durability, 54; driven by mules, 69

Tributes: and population estimates, 9; and population change, 10; importance of, 12; of maize, and lessees, 20; of maize, to mill, 95

Turbine: at mill, 54

Turriano, Juanelo, 54

Tuxtla: plantation, 11; exports of muscovado, 50; imports of slaves, 78, 81; transfer of slaves from, 79, 81, 94

Uncertainty: in sugar production, 40; and oxen numbers, 66–67

United States: acquisition of sugar colonies, 3–4; and races in Mexico, 38; use of mules in, 70

Urtado, Diego, 36, 85

Valley of Mexico: agriculture and population of, 5

Velasco, Luis de: and Indian labor, 87

Vera Cruz, 4

Viceregal edicts: and Indian labor, 86–87; effects on labor inputs, 99–100

Wages: modes of payment, 70; of Spaniards, 75, 77–78; changes in, 78; of naboríos, 89–91; paid for Negro slaves, 99

Wastage: expected loss of sugar through, 61

Water: springs and landforms, 26; supply of Atlacomulco, 42; supply affected by cutting forests, 73

Water need: in Morelos, 40–41; and agricultural scheduling, 42

Water power: abandoned, 63, 69

Water rights: of Tlaltenango, 38; acquisition of, 38–39

Waterwheel: characteristics of, 54

Weeding: use of coa, 45; methods of, 46–47

West Indies: compared with Morelos, 105. *See also* Antilles; British West Indies; French West Indies

Wheat: rations of, 95–96; costs of, 96

Wine: rations of, 97

Wood: replaced by iron, 63–64. *See also* Firewood

Xigo, 79

Xoxo, 79

Yautepec: as site of mills, 5; population changes, 10, 82; silkworms in, 11; and labor supply, 87

Yields: of cane and sugar, 40; compared with West Indies, 45; effect of harvest date on, 46; of sugar, 48–49; changes in, 104–5

Zacatecas, 4

Zacatepec, 4, 25

Zacualpa, 29

Zamba (zambaigo): meaning and importance of, 79

Zapata, Emiliano, 5

Zape, 79

Zumpango: mines, 11

Zúñiga, Juana de: income of, 13, 14